Little
Science,
Big
Science
…and
Beyond

LITTLE
SCIENCE,
BIG
SCIENCE
...AND
BEYOND

DEREK J. DE SOLLA PRICE

COLUMBIA

UNIVERSITY PRESS

NEW YORK 1986

Library of Congress Cataloging-in-Publication Data

Price, Derek J. de Solla (Derek John de Solla), 1922–1983.
Little science, big science—and beyond.

Includes text of: Little science, big science.
Originally published: 1963.
Bibliography: p.
Includes index.
1. Science—Philosophy—Addresses, essays, lectures.
2. Science—Historiography—Addresses, essays, lectures.
3. Science—Social aspects—Addresses, essays, lectures.
I. Price, Derek J. de Solla (Derek John de Solla),
1922–1983. Little Science, big science. 1986.
II. Title.
Q175.3.P73   1986   501   85-19545

ISBN 0-231-04956-0
ISBN 0-231-04957-9 (pbk.)

Columbia University Press
New York   Guildford, Surrey
Copyright © 1986 Columbia University Press
All rights reserved

Printed in the United States of America

This book is Smyth-sewn.

CONTENTS

CONTENTS

Foreword

On its first appearance, this book crystallized a new element in the historiography and sociology of science. It did so in the course of examining the major transformation in the structure of science prefigured in its title: from little to big science. As is often the case with innovative inquiry, the methods of investigation have proved to be rather more consequential for an understanding of the subject than the provisional results reached by use of those methods. For in elucidating the social and cognitive arithmetic of science, this book did much to lay the foundations of the field of inquiry given over to the quantitative analysis of science and scientific development—the field that has come to be known as scientometrics, or, at times, bibliometrics. And although the genealogy of science and learning has become somewhat crowded with the ascribed founders of this or that discipline, of this or that specialty, we can hardly doubt that with this book and the papers which followed it—nine of them included in this new edition—Derek John de Solla Price takes his place as the father of scientometrics.

Throughout the book, its author is mindful of the distant as well as the immediate antecedents of his own approach to the historiography and sociology of science. He invokes the attitudes and practices of that inveterate nineteenth-century measurer of many things, Sir Francis Galton, just as he alludes to Sir William Petty, whose systematic study of bills of mortality in the seventeenth century inaugurated what he described as "political arithmetic." It is symbolically apt, therefore, that in an address to the New York

Academy of Sciences a decade ago, Price should have elected the title "The Political Arithmetic of Science Policy." Himself passionately devoted to the taking of measurements "drawn from many numerical indicators of the various fields and aspects of science," Derek Price can be described, in an almost inevitable eponymous metaphor, as the William-Petty-and-the-Francis-Galton of the historiography and sociology of science.*

Throughout the book, the style of thought and expression bears the author's unmistakable stamp. This is evident from the very first page where we find the pithy aphorism: "we can say that 80 to 90 percent of all the scientists that have ever lived are alive now." So often quoted for so many years, the aphorism in its appearance here may lead some new readers to suppose that Price must himself be quoting, with obliteration of source, from someone's earlier work. But that is only the most familiar case in which he encodes his new ideas in lively and memorable prose. In a felicitous stroke of terminological recoinage, to take another case, he adopts and conceptually extends Robert Boyle's seventeenth-century term, "invisible college," to designate the informal collectives of closely interacting scientists, gener-

---

*Upon reading this foreword, Ellen Price, Derek's Danish-born wife, wrote to provide the ultimate evidence that Derek could put even the great Galton to shame in the depth of his passion for measurement. With her permission, we quote the decisive passage in her letter: "When I began labor with our first child, Linda, in '50, Derek obtained some graph paper to mark down the periodicity of contractions in order to predict the birth-time of the baby—but Nature doesn't work quite like that—and Derek became very angry—God had let him down— darn it, it *ought* to work this way. The conclusion—God was really not very smart—just look at the rotten job he did on optics of the eye!"

A paragraph reminiscent, not least in its beautifully calibrated use of the expressive Sternian dash, of Tristram Shandy's many passages on the vicissitudes of his own birth, while telling everything of Derek's mensurative passion.

ally limited to a size "that can be handled by interpersonal relationships." Invisible colleges, he suggests, are significant social and cognitive formations that advance the research fronts of science, a conception largely confirmed by the early studies of Crane and Mullins and explored in some 300 articles and monographs lately compiled by Chubin.[1] Metaphors such as the invisible college serve to fix in memory some of Price's many contributions to what he describes as "the calculus of science."

Derek Price enjoyed, indeed, actively cultivated, a distinct kind of theoretical panache. In the words of Henry Small, a member of the same invisible college, Price as a theoretician of science took data seriously—but not too seriously. Nor was he given to understatement. Where others might be inclined to speak of "hypotheses" or, at most, of "empirical generalizations," he liked to speak of "laws" of the development of science. No routineer, he created his own orthodoxies but then did not invariably abide by these, either. What did remain intact was a style of thought that could ever after be recognized instantly. His flair for the dramatic often served to call attention to ideas and problems that had long gone unexamined.

Fired by Price's ample numerical imagination, this book is dedicated to establishing and interpreting the magnitudes of growth in "the size of science": in the numbers of scientists and scientific publications and in the societal resources allocated to the pursuit of science and science-based technology. But, as is emphatically asserted, it is not so much the sheer exponential growth in the size of science—an estimated five orders of magnitude in three centuries—as the logistic character of that growth that calls for special notice. It is argued that the inevitable saturation of science will require freshly formulated science policy: "new and exciting tactics for science." Much of the book sets out the Pricean vision of the changing structure and dynamics of scientific work over a wide spectrum ranging from modes of collaboration found in invisible colleges to global aspects of contemporary science.

That vision is enlarged by the array of Price's later papers included in this edition of the book. "Networks of Scientific Papers" is probably his most important single contribution to information science. A pioneering effort to characterize the world network of scientific literature, it indicates that patterns of citation to the papers composing that literature define the parameters of research fronts in science. As the scientometrician Belver Griffith has testified, it was this paper, along with *Little Science, Big Science,* which drew many young scholars, including himself, to the quantitative study of science.

The other papers included here are also innovative. Not since Bertrand Russell had distinguished between "hard" data and "soft" data in his 1914 Lowell Lectures—these being published in the book *Our Knowledge of the External World*—had any historian or sociologist of science undertaken systematic quantitative study of similarities and differences among the various disciplines making diverse use of these types of evidence. In "Citation Measures of Hard Science, Soft Science, Technology, and Nonscience," Price, undaunted by another difficult pioneering effort, undertakes to elucidate certain features that distinguish kinds of scientific from *non*scientific scholarship. This he attempts to do by comparing the proportions of citations in the various disciplines which have high "immediacy" (i.e., references to research published within the preceding five years). He concludes with the hypothesis, still on trial, that the higher the proportion of references to older research in particular works of scholarship, the more probable that they are works of soft science or the humanities.

We refer here to only one more, the last, of the nine papers included in this edition of the book which advance ideas stated or implied in the first edition. Linked with the technique of cocitation analysis introduced by Small and Griffith and with the concept of cumulative advantage in science introduced by Merton,[2] "The Citation Cycle" visualizes an intrinsic structure of cognitive relationships between the scientific archive and newly developing scientific knowledge. A playful endnote tells much-in-little about its author's tiring exchanges with the bureaucracies of science: "This paper acknowledges no support whatsoever from any agency or foundation, but then, no time wasted, either, from preparing and submitting proposals."

Little Science, Big Science has acquired worldwide fame and, much more to the point, has been put to worldwide scholarly use. In light of its author's many-sided applications of citation analysis, it is only apropos to note that the fourteen books he wrote or edited and his approximately 240 published scientific papers have been cited in at least 2,200 articles, a figure that places him well within the highest 1 percent of contemporary cited authors. (That citation figure does not include the unnumbered references to his work in books.) Of all Price's writings, this book has received the greatest notice by far, with some 725 articles referring to it alone. The citations are found in the journals of some 80 disciplines or specialties, ranging from A (aeronautics and anthropology) to Z (zoology) with, of course, the greatest concentration in information science, scientometrics, and the social studies of science. It is drawn upon for its distinctive methods and for the disparate empirical evidence it brings together. Moreover, the book plainly has staying power. The number of references to a scientific paper or book generally peaks about two to five years after publication. In contrast, the references to this book continued to increase for a dozen years and have pretty much maintained that peak plateau during the decade since. One therefore has reason to suppose that the publication of this new edition will lead to a new upswing of attention to it.

The exceptional history of the book led to its being designated as a "Citation Classic" by *Current Contents*, the weekly overview of the contents of scientific and scholarly journals. In accord with the practice of having the author of a citation classic tell how the work came to be, Derek Price wrote his account, shortly before his premature death in 1983. That brief statement can be taken as in effect his preface to this new edition, as we paraphrase his much-quoted aphorism in making a reasonable surmise: most of the future readers of this pathmaking book are probably not yet alive.

Robert K. Merton
Eugene Garfield

Notes

1. Diana Crane, *Invisible Colleges* (Chicago: University of Chicago Press, 1972); Nicholas C. Mullins, *Theories and Theory Groups in Contemporary American Sociology* (New York: Harper and Row, 1973); Daryl E. Chubin, *Sociology of Sciences: An Annotated Bibliography on Invisible Colleges* (New York: Garland, 1983).

2. Henry Small and Belver C. Griffith, "The structure of scientific literatures I: Identifying and graphing specialties," *Science Studies* (1974) 4:17–40; Belver C. Griffith and Henry G. Small, "The structure of scientific literatures II: The

macro- and micro-structure of science," *Science Studies* (1974), 4:339–65; Henry G. Small, "A co-citation model of a scientific specialty: A longitudinal study of collagen research," *Social Studies of Science* (1977), 7:139–66.

Robert K. Merton, "The Normative Structure of Science" [1942], reprinted in Merton, *The Sociology of Science*, Norman W. Storer, ed. (Chicago: University of Chicago Press, 1973), p. 273; "The Matthew effect in science," *Science* (January 5, 1968), 159:56–63, reprinted in Merton, *The Sociology of Science*, pp. 439–59.

Preface to
Little Science, Big Science

Pegram Lecturers are supposed to talk about science and its place in society. The ordinary way of doing this would be either to talk popular science or to adopt one of the various styles in humanistic discussion of the reactions between men and science. Previous lecturers in this series have given accounts of the content of space science and made excursions into the philosophy and the history of science. Although professionally my concern is with the history of science, I have a certain prehistoric past as a physicist, and this had led me to treat these lectures in what is, perhaps, an extraordinary way.

My goal is not discussion of the content of science or even a humanistic analysis of its relations. Rather, I want to clarify these more usual approaches by treating separately all the scientific analyses that may be made of science. Why should we not turn the tools of science on science itself? Why not measure and generalize, make hypotheses, and derive conclusions?

In lectures emanating from so large an atomic establishment as Brookhaven, it would be gratuitous to explain how science has become a crucial and very expensive part of man's activity. In the course of its growth to this condition, science has acquired a great deal of administration, organization, and politicking. These have evolved, for the most part, on an ad hoc, empirical basis. Most of the time I worry that there has been insufficient humanistic appraisal of the situation. In these lectures, I shall worry that we have not been sufficiently scientific in analyzing a whole set of regularities that can be dissected out before beginning to deal humanistically with those irregularities that occur because men are men, and not machines.

My approach will be to deal statistically, in a not very mathematical fashion, with general problems of the shape and size of science and the ground rules governing growth and behavior of science-in-the-large. That is to say, I shall not discuss any part of the detail of scientific discoveries, their use and interrelations. I shall not even discuss specific scientists. Rather, treating science as a measurable entity, I shall attempt to develop a calculus of scientific manpower, literature, talent, and expenditure on a national and on an international scale. From such a calculus I hope to analyze what it is that is essentially new in the present age of Big Science, distinguishing it from the former state of Little Science.

The method to be used is similar to that of thermodynamics, in which is discussed the behavior of a gas under various conditions of temperature and pressure. One does not fix one's gaze on a specific molecule called George, traveling at a specific velocity and being in a specific place at some given instant; one considers only an average of the total assemblage in which some molecules are faster than others, and in which they are spaced out randomly and moving in different directions. On the basis of such an impersonal average, useful things can be said about the behavior of the gas* as a whole, and it is in this way that I want to discuss the analysis of science as a whole.

-----

*One must bear in mind that *gas* derives from the Greek *Khaos,* a perfectly general term for a chaos.

According to this metaphor, my first lecture is concerned with the volume of science, the second with the velocity distribution of its molecules, the third with the way in which the molecules interact with one another, and the fourth in deriving the political and social properties of this gas.

Derek J. de Solla Price

Price's "Citation Classic"

May 18, 1983

In 1949, I was at Raffles College (now the University of Singapore) when their new library, not yet built, received a complete set (1662–1930's) of the *Philosophical Transactions of the Royal Society of London*. I took the beautiful calf-bound volumes into protective custody and set them in ten-year piles on the bedside bookshelves. For a year I read them cover to cover, thereby getting my initial education as a historian of science. As a side product, noting that the piles made a fine exponential curve against the wall, I counted all the other sets of journals I could find and discovered that exponential growth, at an amazingly fast rate, was apparently universal and remarkably long-lived. In 1950, to mark my transition from physics and mathematics to the history of science, and from Asia back to Europe, I gave a paper on the topic to the International Congress for this History of Science in Amsterdam.[1] It passed totally unnoticed, and was very ill-received when I entered Cambridge for a second Ph.D. in the new field. It went over like a lead balloon on a couple more trials, but I included it as the last lecture in an inaugural lecture series when I finally got a chair and a department at Yale University, and it was published in *Science Since Babylon* in 1961.[2]

Although most of my time was then given to straight history of science, mainly in ancient astronomy and scientific instrumentation, the exponential growth business needled me a lot, and I began to pursue other quantitative researches about science, stimulated much by Robert Merton's writings in the sociology of science, by Eugene Garfield's new work on citation indexing,[3] and by rereading

Desmond Bernal's books which had prepared my mind for the initial sensitivity that led me to this field in the first place. A few months after *Science Since Babylon* hit the bookstores, I was asked if I would like to expand that last lecture into a new series for the Pegram lectures at Brookhaven. The series met with an enthusiastic reception from the physicists who were very interactive while I lived there working out the weekly lectures and writing them up for publication as I went along as I had done for the Yale lectures before. I ladled into those lectures all the half-baked results I had collected together in this nonfield over the past several years, and tried to give the whole thing some measure of coherence. It was, apparently, an immediate success, and sold quite well among the scientists, remaining totally alien to the historians and historians of science. *Little Science, Big Science* became a success and a *Citation Classic*, I think, because just at that time there were two new fields emerging as part of the academic explosion of the 1960s, the sociology of science and library science (as distinct from library trade schools). Those two fields seemed to react almost alchemically with my offbeat development of quantitative methods in what was to become science of science or scientometrics; my book was accepted as one of the prime sources for the techniques and results.

Notes

1. Derek J. de Solla Price, "Quantitative measures of the development of science," *Archives Internationales d'Histoire des Sciences* (1951), 14:85–93; "Quantitative measures of the development of science," *Actes du VI Congrès International d'Histoire des Science,* 1950 (Amsterdam, Paris: Herman, 1951), pp. 413–21.

2. Price, *Science Since Babylon* (New Haven: Yale University Press, 1961).

3. Eugene Garfield, *Essays of an Information Scientist,* 5 vols. (Philadelphia, ISI Press, 1983).

Acknowledgments

In the first edition of this book the author acknowledged the following people: his graduate students at Yale; his research assistant, Joy Day; his friend, David Klein; and Asger Aaboe for drawing many of the graphs. He was also indebted to Yale University for granting permission to reproduce figures from *Science Since Babylon;* to the Mc-Graw-Hill Book Company for permission to reproduce the graph on development of accelerators from M.S. Livingston and J.P. Blewett, *Particle Accelerators;* to the Addison Wesley Publishing Company for the graph on growth of cities from George K. Zipf, *Human Behavior and the Principle of Least Effort;* and to Cambridge University Press for permission to use an adaptation of a figure from D'Arcy W. Thompson, *Growth and Form.*

Columbia University Press wishes to thank the following publishers for permission to reprint the articles by Derek Price used in this book. It should be noted that, for the purpose of cross-referencing among essays in the collection, some of the original references and/or footnote formats were changed slightly. It was also necessary to make some minor editorial changes.

"Price's 'Citation Classic,'" printed as "This Week's Citation Classic," *Current Contents* (July 18, 1983), 29:18. Copyright 1983 by the Institute for Scientific Information.

"Networks of Scientific Papers," *Science* (July 30, 1965), 149:510–15. Copyright 1965 by the American Association for the Advancement of Science.

"Collaboration in an Invisible College," *American Psy-*

chologist (November 1966), 21(11):1011–18. Copyright 1966 by the American Psychological Association.

"Measuring the Size of Science," *Proceedings of the Israel Academy of Science and Humanities*, Jerusalem (1969), 4(6):98–111. Paper reprinted in *ISI's Who Is Publishing in Science, 1975 Annual*, pp. 17–25. Copyright 1975 by the Institute for Scientific Information.

"Citation Measures of Hard Science, Soft Science, Technology, and Nonscience," in Carnot E. Nelson and Donald Pollack, eds., *Communication Among Scientists and Engineers*, p. 3–22 (Lexington, Mass.: Lexington Books, D.C. Heath and Company, 1970). Copyright 1970 by D.C. Heath and Company.

"Some Statistical Results for the Numbers of Authors in the States of the United States and the Nations of the World," preface to *ISI's Who Is Publishing in Science 1975 Annual*. Copyright 1975 by the Institute for Scientific Information.

"Studies in Scientometrics, Part 1: Transience and Continuance in Scientific Authorship," *International Forum on Information and Documentation*, International Federation for Documentation, Moscow (1976), 1(2):17–24; also published in *Ciência da Informação*, Rio de Janeiro (1975), 4(1):27–40.

"Studies in Scientometrics, Part 2: The Relation Between Source Author and Cited Author Populations," *International Forum on Information and Documentation*, Moscow (1976), 1(3):19–22; also published in *Ciência de Informação*, Rio de Janeiro (1975), 4(2):103–108.

"Of Sealing Wax and String," *Natural History* (January 1984), 93(1):48–56. Copyright the American Museum of

ACKNOWLEDGMENTS   xxiii

Natural History, 1984; editorial changes approved by Ellen
de Solla Price.

"The Citation Cycle," B.C. Griffith, ed., *Key Papers in
Information Science*, pp. 195–210 (White Plains, N.Y.:
Knowledge Industry Publications, 1980). *Key Papers* was
published by Knowledge Industry Publications, Inc. for the
American Society for Information Science. Copyright 1980
by Knowledge Industry Publications, Inc.

Eugene Garfield, "Price's Citation Cycle," *Current Contents* (September 29, 1980), 39:618–20. Copyright 1980 by
the Institute for Scientific Information.

Little Science, Big Science ...and Beyond

How big are you, baby?

Why, don't you know,

You're only so big,

And there's still room to grow.

(NURSERY RHYME)

1

Prologue to a Science of Science

During a meeting at which a number of great physicists were to give firsthand accounts of their epoch-making discoveries, the chairman opened the proceedings with the remark "Today we are privileged to sit side-by-side with the giants on whose shoulders we stand."[1] This, in a nutshell, exemplifies the peculiar immediacy of science, the recognition that so large a proportion of everything scientific that has ever occurred is happening now, within living memory. To put it another way, using any reasonable definition of a scientist, we can say that 80 to 90 percent of all the scientists that have ever lived are alive now. Alternatively, any young scientist, starting now and looking back at the end of his career upon a normal life span, will find that 80 to 90 percent of all scientific work achieved by the end of the period will have taken place before his very eyes, and that only 10 to 20 percent will antedate his experience.

So strong and dominant a characteristic of science is this immediacy that one finds it at the root of many attitudes taken by scientist and layman toward modern science. It is what makes science seem essentially modern and contemporaneous. As a historian of science, I find myself doing annual battle to justify and uphold the practice of spending more than half our time on the period before Newton, whereas every contemporary scientist around knows that what really counts is science since Einstein.

Because the science we have now so vastly exceeds all that has gone before, we have obviously entered a new age that has been swept clear of all but the basic traditions of the old. Not only are

2   PROLOGUE TO A SCIENCE OF SCIENCE

the manifestations of modern scientific hardware so monumental that they have been usefully compared with the pyramids of Egypt and the great cathedrals of medieval Europe, but the national expenditures of manpower and money on it have suddenly made science a major segment of our national economy. The large-scale character of modern science, new and shining and all-powerful, is so apparent that the happy term "Big Science" has been coined to describe it.[2] Big Science is so new that many of us can remember its beginnings. Big Science is so large that many of us begin to worry about the sheer mass of the monster we have created. Big Science is so different from the former state of affairs that we can look back, perhaps nostalgically, at the Little Science that was once our way of life.

If we are to understand how to live and work in the age newly dawned, it is clearly necessary to appreciate the nature of the transition from Little Science to Big Science. It is only too easy to dramatize the change and see the differences with reckless naïveté. But how much truth is there in the picture of the Little Scientist as the lone, long-haired genius, moldering in an attic or basement workshop, despised by society as a nonconformist, existing in a state of near poverty, motivated by the flame burning within him? And what about the corresponding image of the Big Scientist? Is he honored in Washington, sought after by all the research corporations of the "Boston ring road," part of an elite intellectual brotherhood of co-workers, arbiters of political as well as technological destiny? And the basis of the change—was it an urgent public reaction to the first atomic explosion and the first national shocks of military missiles and satellites? Did it all happen very quickly, with historical roots no deeper in time than the Manhattan Project, Cape Canaveral rocketry, the discovery of penicillin, and the invention of radar and electronic computers?

I think one can give a flat "No" in answer to all these questions.

The images are too naively conceived, and the transition from Little Science to Big Science was less dramatic and more gradual than appears at first. For one thing, it is clear that Little Science contained many elements of the grandiose. And tucked away in some academic corners, modern Big Science probably contains shoestring operations by unknown pioneers who are starting lines of research that will be of decisive interest by 1975. It is the brave exception rather than the rule that key breakthroughs are heralded at birth as important work done by important people.

Historically, there have been numerous big national efforts: the great observatories of Ulugh Beg in Samarkand in the fifteenth century, of Tycho Brahe on his island of Hven in the sixteenth century, and of Jai Singh in India in the seventeenth century, each of which absorbed sensibly large fractions of the available resources of their nations. As international efforts, there were the gigantic expeditions of the eighteenth century to observe the transits of Venus. And, as large-scale hardware, there were the huge electrical machines, produced most notably in Holland in the eighteenth century, machines that in their time seemed to stretch man's scientific engineering to its ultimate capability and to give him the power to manufacture the most extreme physical forces of the universe, rivaling the very lightning and perhaps providing keys to the nature of matter and of life itself. In a way, our dreams for modern accelerators pale by comparison.

But let us not be distracted by history. What shall concern us is not so much the offering of counterexamples to show that Little Science was sometimes big, and Big Science little, but rather a demonstration that such change as has occurred has been remarkably gradual. To get at this we must begin our analysis of science by taking measurements, and in this case it is even more difficult than usual to make such determinations and find out what they mean.

Our starting point will be the empirical statistical evidence drawn from many numerical indicators of the various fields and aspects of science. All of these show with impressive consistency and regularity that if any sufficiently large segment of science is measured in any reasonable way, the normal mode of growth is exponential. That is to say, science grows at compound interest, multiplying by some fixed amount in equal periods of time. Mathematically, the law of exponential growth follows from the simple condition that at any time the rate of growth is proportional to the size of the population or to the total magnitude already achieved—the bigger a thing is, the faster it grows. In this respect it agrees with the common natural law of growth governing the number of human beings in the population of the world or of a particular country, the number of fruit flies growing in a colony in a bottle, or the number of miles of railroad built in the early Industrial Revolution.

It might at first seem as if establishing such an empirical law of growth for science was neither unexpected nor significant. The law has, however, several remarkable features, and from it a number of powerful conclusions can be drawn. Indeed, it is so far-reaching that I have no hesitation in suggesting it as the fundamental law of any analysis of science.

Its most surprising and significant feature is that, unlike most pieces of curve-fitting, the empirical law holds true with high accuracy over long periods of time. Even with a somewhat careless and uncritical choice of the index taken as a measure, one has little trouble in showing that general exponential growth has been maintained for two or three centuries. The law therefore, though at this stage still merely empirical, has a status immediately more significant than the usual short-term economic time series. This leads one to a strong suspicion that the law is more than empirical—and that with suitable definitions of the indices that grow exponentially, one may show, as I later shall, that there is a reasonable theoretical basis for such a law.

A second important feature of the growth of science is that it is surprisingly rapid however it is measured. An exponential increase is best characterized by stating the time required for a doubling in size or for a tenfold increase.[3] Now, depending on what one measures and how, the crude size of science in manpower or in publications tend to double within a period of 10 to 15 years. The 10-year period emerges from those catchall measures that do not distinguish low-grade work from high but adopt a basic, minimal definition of science; the 15-year period results when one is more selective, counting only some more stringent definition of published scientific work and those who produce it. If this stringency is increased so that only scientific work of *very* high quality is counted, then the doubling period is drawn out so that it approaches about 20 years.

The following list shows the order of magnitudes of an assortment of measurable and estimatable doubling times and shows how rapidly the growth of science and technology has been outstripping that of the size of the population and of our nonscientific institutions.

100 years
    Entries in dictionaries of national biography
50 years
    Labor force
    Population
    Number of universities
20 years
    Gross National Product
    Important discoveries
    Important physicists
    Number of chemical elements known
    Accuracy of instruments
    College entrants/1000 population

15 years
    B.A., B.Sc.
    Scientific journals
    Membership of scientific institutes
    Number of chemical compounds known
    Number of scientific abstracts, all fields
10 years
    Number of asteroids known
    Literature in theory of determinants
    Literature in non-Euclidean geometry
    Literature in X rays
    Literature in experimental psychology
    Number of telephones in United States
    Number of engineers in United States
    Speed of transportation
    Kilowatt-hours of electricity
5 years
    Number of overseas telephone calls
    Magnetic permeability of iron
1½ years
    Million electron volts of accelerators

Bearing in mind the long period of validity of exponential growth, let us note that a 15-year doubling time extended over three centuries of growth corresponds to an increase of 20 powers of two, or a factor of about one million. Thus, in the interval from 1660 to the present day, such indices of the size of science should have increased by the order of a million. To offer the soundest explanation of the scientific and industrial revolutions is to posit that this is indeed what has been happening.

Just after 1660, the first national scientific societies in the modern tradition were founded; they established the first scientific periodicals, and scientists found themselves beginning to write scientific papers instead of the books that hitherto had been their only outlets. We have now a world list of some 50,000 scientific periodicals (fig. 1.1) that have been founded, of which about

30,000 are still being published; these have produced a world total of about six million scientific papers (fig. 1.2) and an increase at the approximate rate of at least half a million a year.[4] In general, the same applies to scientific manpower. Whereas in the mid-seventeenth century there were a few scientific men—a denumerable few who were countable and namable—there is now in the United States alone a population on the order of a million with scientific and technical degrees (fig. 1.3). What is more, the same exponential law accounts quite well for all the time in between. The present million came through intermediate stages of 100,000 in 1900, 10,000 in 1850, and 1000 in 1800. In terms of magnitude alone, the transition from Little Science to Big Science has been steady—or at least has had only minor periodic fluctuations similar to those of the stock market—and it has followed a law of exponential growth with the time rates previously stated.

Thus, the steady doubling every 15 years or so that has brought us into the present scientific age has produced the peculiar immediacy that enables us to say that so much of science is current and that so many of its practitioners are alive. If we start with the law that the number of living scientists doubles in, let us say, 15 years, then in any interval of 15 years there will come into being as many scientists again as in the whole of time preceding. But at any moment there coexists a body of scientists produced not over 15 years but over an interval nearer to the 45 years separating average date of arrival at the research front from average date of retirement from active scientific work. Thus, for every one person born before such a period of 45 years, we now have one born in the first doubling period, two in the second, and four in the third. There are, then, about seven scientists alive for every eight that have ever been, a fraction of 87½ percent; let us call this a coefficient of immediacy. One may calculate this exactly by using actuarial mortality tables, but in fact the result is not much altered

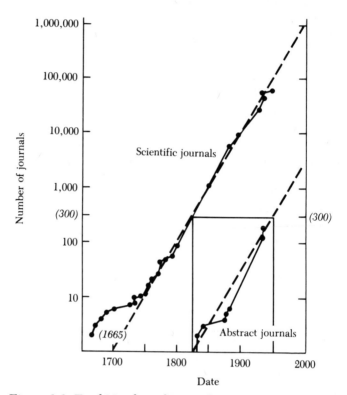

Figure 1.1. Total Number of Scientific Journals and Abstract
Journals Founded, as a Function of Date

Note that abstracts begin when the population of journals is approximately 300.
Numbers recorded here are for journals founded, rather than those surviving; for
all periodicals containing any "science" rather than for "strictly scientific" journals.
Tighter definitions might reduce the absolute numbers by an order of magnitude,
but the general trend remains constant for all definitions. From Derek J. de Solla
Price, *Science Since Babylon* (New Haven: Yale University Press, 1961).

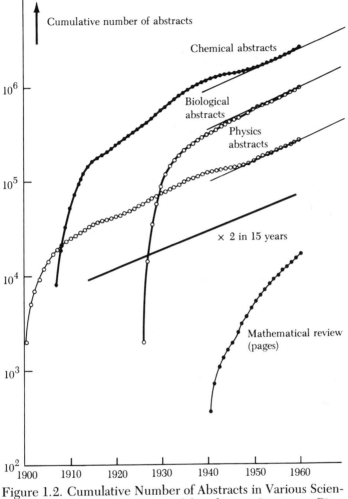

Figure 1.2. Cumulative Number of Abstracts in Various Scientific Fields, from the Beginning of the Abstract Service to Given Date

It will be noted that after an initial period of rapid expansion to a stable growth rate, the number of abstracts increases exponentially, doubling in approximately 15 years.

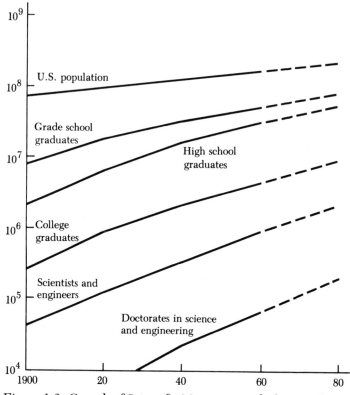

Figure 1.3. Growth of Scientific Manpower and of General Population in the United States

It may be seen that the more highly qualified the manpower, the greater has been its growth rate. It will also be noted that there appears a distinct tendency for the curves to turn toward a ceiling value running parallel with the population curve.

by this because the doubling period of science is so much less than the average working life of a scientist.

For a doubling period of 10 years, the corresponding coefficient of immediacy is about 96 percent; for a doubling time of 20 years, about 81 percent. Thus, even if one admits only the general form of the growth function and the order of magnitude of its time constant, these account for the feeling that most of the great scientists are still with us, and that the greater part of scientific work has been produced within living memory, within the span of the present generation of scientists. Furthermore, one can emphasize the principle by remarking that some time between the next decade and the one after we shall have produced as much scientific work and as many scientists as in the whole time up to the present.

What I have said so far is by now well known and reasonably well agreed upon by those who speculate about science for fun or high policy. I should like to extend these results, however, in a couple of ways that may suggest that this outlook requires revision. In the first place, speaking in terms of a "coefficient of immediacy" can be misleading. Let us compare the figures just found with the conjectural figures for world population.

At the beginning of the Christian era, the human race numbered about 250 million; it grew slowly and erratically, differently in different places and at different times, and reached a figure of 550 million by the mid–seventeenth century. It has grown at an ever-increasing pace, so that today there are about 3000 million people, and it looks as though that number will double every 40 to 50 years. If we reckon about 20 years to a generation, there must have been at least 60,000 millions of people, and thus only about 5 percent of those who have lived since the beginning of our era are alive now. If we count all those who lived before the time of Christ, the fraction will be smaller; if we count only those who have lived since the mid–seventeenth century, it will be a little more than 10 percent. Making due allowance for changing mortalities and age of childbearing will not, I feel, materially alter the qualitative result that the human population is far from immediate in the sense that science is.

Even if we accept the gloomy prognostications of those who talk about the admittedly serious problem of the population explosion, it would apparently take about another half-century— some time after the year 2000—before we could claim that 50 percent of all the human beings that have lived were at that moment alive. Most of the persons that have ever lived are *dead*, and, in the sense that this will continue to be so, they will stay dead. One might conclude, since the rate of growth of entries in the great dictionaries of national biographies shows a fairly constant proportion to the population at various dates, that most of the great or worthy persons of the world are dead. That is why *history* is a subject rather different from *history of science*. There is much more past to live in if you discuss politics and wars than if you discuss science.

The immediacy of science needs a comparison of this sort before one can realize that it implies an explosion of science dwarfing that of the population, and indeed all other explosions of nonscientific human growth. Roughly speaking, every doubling of the population has produced at least three doublings of the number of scientists, so that the size of science is eight times what it was and the number of scientists per million population has multiplied by four. Mankind's per capita involvement with science has thus been growing much more rapidly than the population.

A second clarification, one of crucial importance, must be made concerning the immediacy and growth of modern science. We have already shown that the 80- to 90-percent currency of modern science is a direct result of an exponential growth that has been steady and consistent for a long time. It follows that this result, true now, must also have been true *at all times in the past*, back to the eighteenth century and perhaps even as far back as the late seventeenth. In 1900, in 1800, and perhaps in 1700, one could look back and say that most of the scientists that have ever been are alive now, and most of what is known has been determined within living memory. In that respect, surprised though we may be to find it so, the scientific world is no different now from what it has always been since the seventeenth century. Science has always been modern; it has always been exploding into the population, always on the brink of its expansive revolution. Scientists have always felt themselves to be awash in a sea of scientific literature that augments in each decade as much as in all times before.

It is not difficult to find good historical authority for this feeling in all epochs. In the nineteenth century we have Charles Babbage in England and Nathaniel Bowditch in the United States bitterly deploring the lack of recognition of the new scientific era that had just burst upon them. In the eighteenth century there were the first furtive moves toward special journals and abstracts in a vain attempt to halt or at least rationalize the rising tide of publications; there was Sir Humphrey Davy, whose habit it was to throw books away after reading on the principle that no man could ever have the time or occasion to read the same thing twice. Even in the seventeenth century, we must not forget that the motivating purpose of the *Philosophical Transactions of the Royal Society* and the *Journal des Sçavans* was *not* the publishing of new scientific papers so much as the monitoring and digesting of the learned publications and letters that now were too much for one man to cope with in his daily reading and correspondence.[5]

The principle of more than 80 percent being contemporaneous is clearly sufficient to cast out any naive idea that sheer change in scale has led us from Little Science to Big Science. If we are to distinguish the present phase as something new, something different from the perception of a burgeoning science that was common to Maxwell, to Franklin, and to Newton, then we cannot rest our case on the rate of growth alone. A science that has advanced steadily through more than five orders of magnitude in more than 250 years is not going to be upset by a mere additional single order of magnitude such as we have experienced within the last few decades of the present century.

As a side point one may note that the constancy of this phenomenon of immediacy is typical of many other constancies in science that make it meaningful and useful to pursue the history of science even though most of our past is alive. What we must do in the humanistic and the scientific analyses of science is search out such constancies of scientific method, of public reaction, of the use of mathematical models or euphoric hardware or the ground rules of manpower and motivation, and apply them to our criticism and understanding of this science that seems so essentially modern and out of all relation to Archimedes or Galileo or Boyle or Benjamin Franklin. If we honor a Boyle for his law, or a Planck for his constant, this is largely accidental hero worship; more important to us than the names of those who have quarried a slab of immortality is their having done so in a manner which notably illustrates the constant and seemingly eternal way in which these things have been going on. To take an early example such as Galileo, seen in all its historical perspective, is in many ways more efficient than choosing a recent example such as Oppenheimer, though Galileo can tell us nothing of the content of modern atomic physics as can Oppenheimer.

To return to my main point, if the sheer growth of science in its exponential climb is *not* admissible as an explanation for the transition from Little Science to Big Science, we are left in a quandary. To escape from it one may be tempted at first to deny that there has been any such radical transformation of the state of science. This is amply belied by the fact that since World War II we have been worried about questions of scientific manpower and literature, government spending, and military power in ways that seem quite different, not merely in scale, from all that went before.

Even if one admits that new things are happening and that Big Science differs not merely in scale from Little Science, one might still maintain that it was the cataclysmic changes associated with World War II that initiated us into the new era and produced all the major changes. Quite unexpectedly, one can show from the statistical studies I have been using to measure the pure growth that the influence of the war on scientific manpower and literature seems only to have been the production of a temporary perturbation that extended for its duration.

For this interval it is not possible to use the indices one might use before or after; manpower may be in military service, publication may be suppressed for secrecy. Yet it is apparent that the exponential increase after the war is identical with that before (fig. 1.4). This is a strong result, for it shows that the percentage increase per annum is the same before and after the war and, therefore, if there is any constancy about the way in which scientific papers generate new scientific papers and researchers generate new classes of researchers, there cannot have been any great loss or gain to science during the war. With the exception only of a sidewise displacement of the curve due to secrecy loss, science is just where it would have been, statistically speaking, and is growing at the same rate as if there had been no war. The order of events might have been different, the political implications perhaps grossly so, but there is some reason for taking a fatalistic line that it was in the nature of things for accelerator laboratories to grow as large as Brookhaven, and missile establishments as large as Cape Canaveral, and that had there been no Manhattan Project there might still have been a Sputnik. The war looms as a huge milepost, but it stands at the side of a straight road of exponential growth.

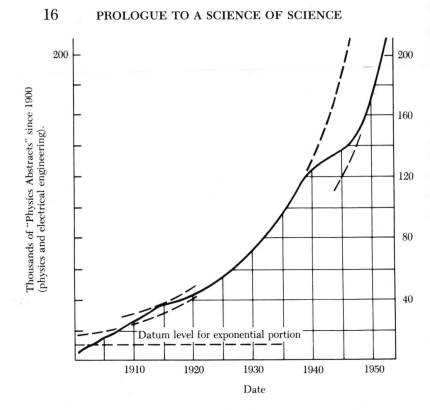

Figure 1.4. Total Number of Physics Abstracts Published Since
January 1, 1900

The full curve gives the total, and the broken curve represents the exponential
approximation. Parallel curves are drawn to enable the effect of the wars to be
illustrated. From Derek J. de Solla Price, *Science Since Babylon* (New Haven;
Yale University Press, 1961).

If, then, we are to analyze the peculiarities of Big Science, we
must search for whatever there is other than the steady, hand-in-
hand climb of all the indices of science through successive orders
of magnitude. There are, I propose, two quite different types of
general statistical phenomena of science-in-the-large. On the one

hand, although we have the overall picture of a steady exponential growth with this amazingly short time constant of about 15 years, not all things are growing at precisely this rate; some are faster, others slower, though all of them outpace the growth of the population. On the other hand, we have the possibility that the exponential law of growth may be beginning to break down.

It is just possible that the tradition of more than 250 years represents a sort of adolescent stage during which every half-century science grew out of its order of magnitude, donned a new suit of clothes, and was ready to expand again. Perhaps now a postadolescent quiescence has set in, and such exuberant growth has slowed down and is about to stop upon the attainment of adult stature. After all, five orders of magnitude is rather a lot. Scientists and engineers are now a couple of percent of the labor force of the United States, and the annual expenditure on research and development is about the same fraction of the Gross National Product. It is clear that we cannot go up another two orders of magnitude as we have climbed the last five. If we did, we should have two scientists for every man, woman, child, and dog in the population, and we should spend on them twice as much money as we had. Scientific doomsday is therefore less than a century distant.

At a later point I shall treat separately the problem of growths at rates different from that of basic exponential increase. We shall consider such growths as slowly changing statistical distributions of the indices rather than as separate rates of increase. Thus, for example, if the number of science Ph.D.'s were doubling every 15 years, and the number of good ones only every 20 years, the quota of Ph.D.'s per good physicist would be doubling only every 60 years, a change so slow that we can count it out of the scientific explosion. I shall show also, from the statistical distribution, that it is reasonable on theoretical grounds to suppose that the doubling time of one measure might be a multiple of the period for some other index. This treatment, however, requires a closer look at what is actually being measured and must be deferred until further results have been achieved from the study of the crude shape of exponential growth.

Moreover, the "normal" law of growth that we have been considering thus far describes, in fact, a most abnormal state of events. In the real world things do not grow and grow until they reach infinity. Rather, exponential growth eventually reaches some limit, at which the process must slacken and stop before reaching absurdity. This more realistic function is also well known as the logistic curve, and it exists in several slightly different mathematical forms. Again, at this stage of ignorance of science in analysis, we are not particularly concerned with the detailed mathematics or precise formulation of measurements. For the first approximation (or, more accurately, the zeroth-order approximation) let it suffice to consider the general trend of the growth.

The logistic curve is limited by a floor—that is, by the base value of the index of growth, usually zero—and by a ceiling, which is the ultimate value of the growth beyond which it cannot go in its accustomed fashion (fig. 1.5). In its typical pattern, growth starts exponentially and maintains this pace to a point almost halfway between floor and ceiling, where it has an inflection. After this, the pace of growth declines so that the curve continues toward the ceiling in a manner symmetrical with the way in which it climbed from the floor to the midpoint. This symmetry is an interesting property; rarely in nature does one find asymmetrical logistic curves that use up one more parameter to describe them. Nature appears to be parsimonious with her parameters of growth.

Because of the symmetry so often found in the logistic curves that describe the growth of organisms, natural and manmade,

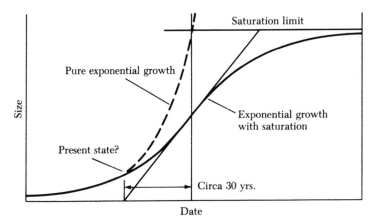

Figure 1.5. General Form of the Logistic Curve

From Derek J. de Solla Price, *Science Since Babylon* (New Haven, Yale University Press, 1961).

measuring science or measuring the number of fruit flies in a bottle, the width of the curve can be simply defined. Mathematically, of course, the curve extends to infinity in both directions along the time axis. For convenience we measure the width of the midregion cut off by the tangent at the point of inflection, a quantity corresponding to the distance between the quartiles on a standard curve of error or its integral. This midregion may be shown necessarily to extend on either side of the center for a distance equal to about three of the doubling periods of the exponential growth.

Thus, for example, if we have a beanstalk that doubles in height every day, there will exist a midperiod of about six days during which the beanstalk will leave its juvenile exponential growth and settle down to an adult life of stability in length (fig. 1.6). The only question is one of how much free and exponential growth is allowed before the decelerative period sets in. For the beanstalk, the midpoint of growth occurs only about four days after the onset of the process, so that there is but one day of relatively free growth, and final length is attained after seven days. Note that the analysis involved no knowledge about the height of the curve from floor to ceiling. True, we made a statement about the date of the midpoint—it occurred after four days of growth—but we could equally well have noted that the exponential growth, short-lived in this case, extends only for the first day, and from this it would follow that three more doublings must bring it to the midpoint, and a further three to senescence.

Now, with no stronger assumption than has been made about the previously regular exponential growth with a doubling period

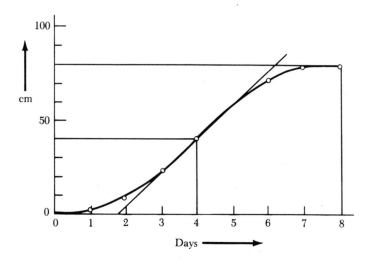

Figure 1.6. Growth in Length of a Beanstalk as a Function of Age

Adapted from D'Arcy W. Thompson, *Growth and Form* (Cambridge: Cambridge University Press, 1948), p. 116, figure 20.

of 10 to 15 years, we may deduce, as we have, that the existence of a ceiling is plausible since we should otherwise reach absurd conditions at the end of another century. Given the existence of such a limit, we must conclude that our exponential growth is merely the beginning of a logistic curve in other guise. Moreover, it is seen that as soon as one enters the midregion near the inflection—that period of secession from accustomed conditions of exponential growth—then another 30 to 45 years will elapse before the exact midpoint between floor and ceiling is reached. An equal period thereafter, the curve will effectively have reached its limit. Thus, without reference to the present state of affairs or any estimate of just when and where the ceiling is to be imposed, it is apparent that over a period of one human generation science will suffer a loss of its traditional exponential growth and approach the critical point marking its senile limit.

However, growths that have long been exponential seem not to relish the idea of being flattened. Before they reach a midpoint they begin to twist and turn, and, like impish spirits, change their shapes and definitions so as not to be exterminated against that terrible ceiling (fig. 1.7). Or, in less anthropomorphic terms, the cybernetic phenomenon of hunting sets in and the curve begins to oscillate wildly. The newly felt constriction produces restorative reaction, but the restored growth first wildly overshoots the mark and then plunges to greater depths than before. If the reaction is successful, its value usually seems to lie in so transforming what is being measured that it takes a new lease on life and rises with a new vigor until, at last, it must meet its doom.

One therefore finds two variants of the traditional logistic curve that are more frequent than the plain S-shaped ogive. In both cases the variant sets in some time during the inflection, presumably at a time when the privations of the loss of exponential growth become unbearable. If a slight change of definition of the thing

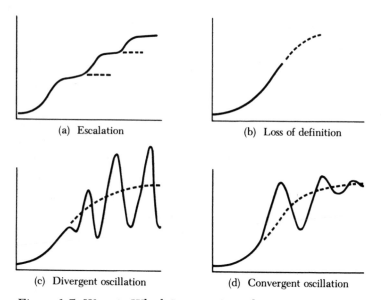

(a) Escalation

(b) Loss of definition

(c) Divergent oscillation

(d) Convergent oscillation

Figure 1.7. Ways in Which Logistic Growth May React to Ceiling Conditions

In escalation, new logistics are born as the old ones die, in loss of definition it becomes impossible to continue to measure the variable in the same way or in the same units, and in oscillation (convergent and divergent) cybernetic forces attempt to restore free growth.

that is being measured can be so allowed as to count a new phenomenon on equal terms with the old, the new logistic curve rises phoenixlike on the ashes of the old, a phenomenon first adequately recognized by Holton and felicitously called by him "escalation." Alternatively, if the changed conditions do not admit a new exponential growth, there will be violent fluctuations persisting until the statistic becomes so ill-defined as to be uncountable, or in some cases the fluctuations decline logarithmically to a stable maximum. At times death may even follow this attainment of maturity, so that instead of a stable maximum there is a slow decline back to zero, or a sudden change of definition making it impossible to measure the index and terminating the curve abruptly in midair.

Logistic curves such as these have become well known in numerous analyses of historical time series, especially those concerning the growth of science and technology. The plain curve is well illustrated in the birth and death of railroad track mileage; in this case the maximum is followed by an eventual decline as tracks are torn up and lines closed down. The curve followed by hunting fluctuations appears in the figures for the production of such technological raw materials as coal and metals (fig. 1.8).[6] The escalated curves are probably the most common and can be seen in the number of universities founded; the separate steps here beautifully reflect the different traditions of the medieval universities and the Renaissance foundations (fig. 1.9).

They can be seen again in the now familiar graph, first presented humorously by Fermi,[7] showing the power of accelerators (fig. 1.10). It becomes less and less humorous as it goes on faithfully predicting when yet another major advance in method is needed to produce another step in the escalation. Yet, again, escalations can be seen in the curve showing the number of chemical elements known as a function of date (fig. 1.11). Omitting the first ten, which were known to prehistoric man, we have a steady exponential growth, doubling a shade more rapidly than every 20 years, followed by a midpoint in about 1807 when Sir Humphrey Davy had his heyday, then a period of decline when the first 60 elements had been found. By the end of the nineteenth century, when new methods, physical rather than chemical, led to new classes of elements, there appeared a new bunch of ogives, then a halt until the big machines enabled man to create the last batch of highly unstable and short-lived transuranic elements.

From this we are led to suggest a second basic law of the analysis

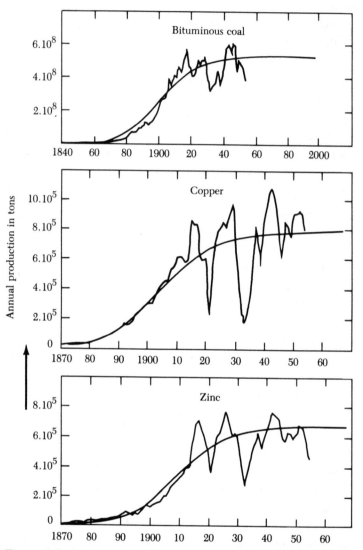

Figure 1.8. Logistic Growth of Raw Material Production, Showing Oscillation on Attaining Ceiling Conditons

Adapted from S. G. Lasky, "Mineral industry futures can be predicted," *Engineering and Mining Journal* (September 1955), Vol. 15b.

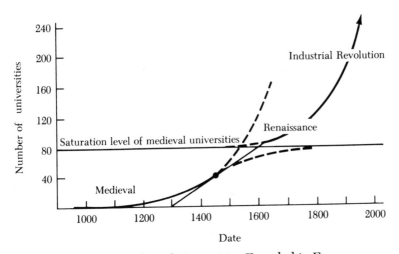

Figure 1.9. Number of Universities Founded in Europe

From the foundation at Cairo in 950 up to ca. 1460 there is pure exponential growth, doubling in about 100 years. Thereafter saturation sets in, so that the midregion of the sigmoid extends from 1300 to ca. 1610. Between 1460 and 1610 is a period of transition to the new form of universities, a growth that also proceeds exponentially as if it had started from unity ca. 1450 and doubling every 66 years. There is probably an ever-greater transition to yet faster growth starting at the end of the Industrial Revolution. From Derek J. de Solla Price, *Science Since Babylon* (New Haven: Yale University Press, 1961).

of science: all the apparently exponential laws of growth must ultimately be logistic, and this implies a period of crisis extending on either side of the date of midpoint for about a generation. The outcome of the battle at the point of no return is complete reorganization or violent fluctuation or death of the variable.

Now that we know something about the pathological after-life of a logistic curve, and that such things occur in practice in several special branches of science and technology, let us reopen the question of the growth curve of science as a whole. We have seen that it has had an extraordinarily long life of purely exponential

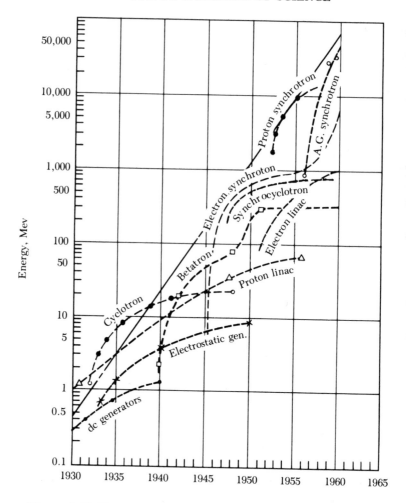

Figure 1.10. The Rate of Increase of Operating Energy in Particle Accelerators

From M. S. Livingston and J. P. Blewett, *Particle Accelerators* (New York: McGraw-Hill, 1962), p. 6, figure 1.1, used by permission.

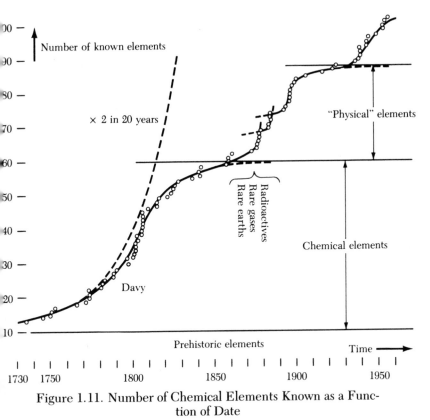

Figure 1.11. Number of Chemical Elements Known as a Function of Date

After the work of Davy there is a clear logistic decline followed by a set of escalations corresponding to the discovery of elements by techniques that are predominantly physical. Around 1950 is the latest escalation produced by the manufacture of transuranic elements.

growth and that at some time this must begin to break down and be followed by a generation-long interval of increasing restraint which may tauten its sinews for a jump either toward escalation or toward violent fluctuation. The detailed nature of this change, and any interpretation of it, must depend on what we are measuring and on how such an index is compiled.

Even without such definition and analysis one can immediately deduce various characteristics of such a period. Clearly there will be rapidly increasing concern over those problems of manpower, literature, and expenditure that demand solution by reorganization. Further, such changes as are successful will lead to a fresh escalation of rapid adaptation and growth. Changes not efficient or radical enough to cause such an offshoot will lead to a hunting, producing violent fluctuations that will perhaps smooth out at last.

Such an analysis seems to imply that the state called Big Science actually marks the onset of those new conditons that will break the tradition of centuries and give rise to new escalations, violent huntings, redefinitions of our basic terms, and all the other phenomena associated with the upper limit. I will suggest that at some time, undetermined as yet but probably during the 1940s or 1950s, we passed through the midperiod in general logistic growth of science's body politic.

Thus, although we recognize from our discussion so far that saturation is ultimately inevitable, it is far too approximate to indicate when and in what circumstances saturation will begin. I now maintain that it may already have arrived. It may seem odd to suggest this when we have used only a few percent of the manpower and money of the country, but in the next chapter it will appear that this few percent actually represents an approach to saturation and an exhausting of our resources that nearly (within a factor of two) scrapes the bottom of the barrel.

At all events, the appearance of new phenomena in the involvement of science with society seems to indicate something radically different from the steady growth characteristic of the entire historic past. The new era shows all the familiar syndromes of saturation. This, I must add, is a counsel of hope rather than despair.

Saturation seldom implies death, but rather that we have the beginning of new and exciting tactics for science, operating with quite new ground rules.

It is, however, a grave business, for Big Science interpreted thus becomes an uncomfortably brief interlude between the traditional centuries of Little Science and the impending period following transition. If we expect to discourse in scientific style about science, and to plan accordingly, we shall have to call this approaching period New Science, or Stable Saturation; if we have no such hopes, we must call it senility.

2

Galton Revisited

Francis Galton (1822–1911), grandson of Erasmus Darwin, was one of the most versatile and curious minds of the nineteenth century. He brought fingerprinting to Scotland Yard, founded the Eugenic Society which advocated breeding of the human race on rational principles, and, above all, gave a flying start to the science of mathematical genetics. His passion was to count everything and reduce it to statistics. Those who see the social sciences rising on a solid foundation of quantified measurements and mathematical theory might well take him as a patron saint rather than Sir William Petty, who is usually seen as the first to bring numbers into the study of people by analyzing the bills of mortality in the seventeenth century.

Galton's passion shows itself best, I feel, in two essays that may seem more frivolous to us than they did to him. In the first, he computed the additional years of life enjoyed by the Royal Family and the clergy because of the prayers offered up for them by the greater part of the population; the result was a negative number. In the second, to relieve the tedium of sitting for a portrait painter, on two different occasions he computed the number of brush strokes and found about 20,000 to the portrait—just the same number, he calculated, as the hand movements that went into the knitting of a pair of socks.[1]

Let it not be thought that Galton was some sort of crank. His serious work was of the highest standards of scholarship and importance, but he is now increasingly neglected because, although his researches were founded on the exciting and valid basis of Darwin's theory of evolution, Galton had missed the true mechanism of genetic action, discovered by his exact contemporary Mendel. Mendel published his findings just five years before Galton's book on hereditary genius,[2] but was not discovered by the outside world until Galton was nearly eighty.

We shall examine his book *Hereditary Genius*, and, with particular attention, his special study, *English Men of Science* (London 1874). In these works Galton is primarily concerned with his thesis that great men, including creative scientists, tend to be related and that therefore a series of elite families contributed perhaps the majority of distinguished statesmen, scientists, poets, judges, and military commanders, of his day and of the past. His main work is full of pitfalls, and currently we are not concerned so much with the Galtonian approach to genetics as we are with several of his interesting side investigations. These are his pioneer studies of the distribution of quality among distinguished scientists, and a set of summaries that we should nowadays call sociological and psychological, telling us something about the characteristics of these exceptional men.

We intend to review these two main lines in the light of the twentieth century and its extensions of Galton's work. The first, telling us how many men or scientific papers or pieces of research there are at each of several levels of quality, is necessary if we are to understand the nature of scientific quality, and this knowledge is a prerequisite to the interpretation of the several different index measures previously mentioned in connection with the basic laws governing the exponential and logistic rates of growth of science. The second will help us formulate ground rules for what to expect of scientists when the change of conditions produced by Big Science or Saturation Science alters their circumstances from those they had known in past ages.

Galton began by estimating how rare in the England of his day were various types of men who were engaged in human affairs generally and in science particularly and who were of sundry degrees of eminence. Using the criterion that a man was eminent if his name appeared in a short biographical compilation of 2500 *Men of the Time* that had just been published, or in the select columns of obituary notices in *The Times,* he found that such noteworthiness had an incidence of about one person for every 2000 adult males or one person in 20,000 of the general population—a mere handful alive at any time in the country.

For eminent scientists he set a standard which demanded that they should be not merely Fellows of the Royal Society—a meaningful honor since the reforms of election under Mr. Justice Groves some thirty years before—but that they must be further distinguished by a university chair; by a medal presented by a learned society, or an office held in such a body; or by membership in some elite scientific club of academic worth. His count of people from whom he could obtain the full biographical information desired was 180, and he estimated that in the entire country there might be at the most 300 such people.

Reckoning that half of them were between the ages of fifty and sixty-five, he calculated that the chance of rising to such stature was about 1 in 10,000 adult males of this age group, a figure roughly corresponding to 1 in 100,000 of the general population. However, since the general biographical lists show that only about 1 in 10 eminent men was engaged in science or medicine, then by his previous standards there should only have been about one eminent scientist for every 200,000 of the general population. The fact that Galton supposes there to have been twice as many, means either that he was erring on the side of generosity in estimating the numbers of good scientists who should have been on his list and were not, or that the tendency is to cast a broader net when looking for great scientists than when looking for great men in general.

The utility of this investigation is that it provides an estimate of the number of scientific persons whom Galton considered important enough to be well worth discussion, but without limiting the scope to include such a small group that it would leave the investigator generalizing about a mere handful of geniuses. Thus, between 5 and 10 persons in a million fall within this category. How does this compare with the state of affairs since Galton's time?

Fortunately there is an admirable biographical compilation, *American Men of Science,* that has run through ten editions between 1903 and 1960. The editor, J. McKeen Cattell (himself a prominent psychologist), rendered signal service by starring the most noteworthy names, beginning with an original 1000 and adding to this number as each new edition appeared.[3] It so happens that in the first edition there are about 11 starred names to a million population of the United States, in the volume for 1938 there are about 12.4 to a million, and that both figures are of the same order of magnitude as that found by Galton. Certainly there appears to have been no vast change in the number of "eminent" men of science to a million population, either on moving our scene of inquiry from Britain to the United States or on following it through nearly a doubling of the United States population. One may argue that Galton's standard of distinction is not the same as Cattell's. One may maintain with even greater reason that Cattell's arbitrary allocation of a set quota of 1000 stars originally, with a fixed increment thereafter, was perhaps out of all proportion to a constant standard of eminence. In spite of this, we can find no rapid changes in this estimated incidence of scientific eminence.

If in studying *American Men of Science* we look not at the starred names alone but at all of them, we observe a most striking change in order of magnitude with the passing of time (table 2.1). Just to run one's eye along the set of ten editions on a shelf is to feel an immediate respect for the power of exponential growth.

It is apparent that within the past 50 years there has been a sixteenfold increase in the number of men, an exponential growth with a doubling period of about 12½ years, a figure already suggested as typifying the growth of science. Even in relation to the size of the general population it can be seen that the same half-century has multiplied the density of scientists by a factor of 8, a doubling in about 17 years. Another four such half-centuries of regular growth would give us more than two million American men of science per million population, if it were not that exponential growths inevitably become logistic and die.

We have already shown that because of this logistic machinery the prospect for the immediate future is more interesting than that of a slow death from suffocation in a.d. 2160. Our crisis seems

Table 2.1
Number of Men Cited in Editions of *American Men of Science*

Year of Publication	Number of Men	Number per Million Population U.S.
1903	4,000	50
1910	5,500	60
1921	9,500	90
1928	13,500	110
1933	22,000	175
1938	28,000	220
1944	34,000	240
1948	50,000	340
1955	74,000	440
1960 (omitting social sciences)	96,000	480

to be but a few decades ahead, and far more involved with the nature of the growth than with the final exhaustion of the population. It is therefore a matter of some interest to seek the reason why, in spite of this general rapid exponential growth of scientific manpower—and, incidentally, its publications and budgets—the number of truly great men does not seem to change with the same quick exuberance.

The root of the trouble, as Galton well perceived, lies in the establishment of any objective standard of eminence not dependent upon time. Conceivably, all that we have said is that when men are chosen by degrees of selectivity that run to orders of 10 in a million they become remarkable to this standard extent. The same problem is encountered in most recent evaluations of the high-talent population on the basis of intelligence tests. For example, one may say that on an AGCT (Army General Classification Test) type of test only one man in 10,000 of his age group might score more than 170, one in 100,000 more than 180, one in 1,000,000 more than 190, and so on, with an order of magnitude for each 10 (more accurately, 11) points that raise the stakes. But one cannot usefully say that eminence begins at a score of 172 and not below. Even if genius were merely a matter of the talents being measured by the test in hand, there would be no clear cutoff, only a gradual falling off of the population as the standards are raised. The fault lies not so much in the definition of what constitutes scientific ability as in the false premise that distinction or genius can be decided on a yes-or-no-basis.

Results more accurate, although not much more, can be achieved by taking a reasonably small group of tabulated men, discoveries, or even scientific institutions, journals, and countries, and carefully marking in some special way those that were distinguished. For eminent men, for example, one might use as criteria selection to give invited papers, and to receive medals and other awards, such as Nobel Prizes. This gives the usual sort of exponential growth, but with a doubling time considerably longer than 10 years. For example, in a select, apparently superior group of modern scientists in any large field, drawn from standard biographical handbooks or other sources that select only a small elite, the doubling time is about 20 years. One obtains about the same figure for any list of selected great scientific discoveries.

To improve the strength and significance of this result, it is clearly necessary to make some statement about degrees of eminence that would give not a dichotomy of distinguished and undistinguished but rather a sliding scale, a sort of velocity distribution. One such scale—the traditional one used by deans and other employers as a measure of scientific success—is the number of publications produced by each man in accepted scientific journals. Let it be freely admitted at the outset that this is a bad scale. Who dares to balance one paper of Einstein on relativity against even a hundred papers by John Doe, Ph.D., on the elastic constant of the various timbers (one to a paper) of the forests of Lower Basutoland?

The scale is bad if for no other reason than that its existence has moved people to publish merely because this is how they may be judged. Nevertheless, it makes a starting point, and later on it may be refined to meet objections. We shall show, for example, that all such distributions are of the same type and, thus, though one cannot directly measure "scientific ability," one may reasonably deduce properties of its presumed distribution. We shall also have to enter the caveat that the scale may not be directly applicable to the era of Big Science, which has involved so much collaborative work that one cannot easily determine a man's score. This is another point to be reserved for later elaboration.

Let us not begin with too pessimistic an outlook on the worth of this investigation. Flagrant violations there may be, but on the whole there is, whether we like it or not, a reasonably good correlation between the eminence of a scientist and his productivity of papers. It takes persistence and perseverance to be a good scientist, and these are frequently reflected in a sustained production of scholarly writing. Then, again, it may be well demonstrated that the list of high scorers contains a large proportion of names that are not only well known but even honored. Conversely, the low-scoring end of the list contains fewer such names in terms of absolute numbers, and much fewer in proportion.

Exactly such a study was made by Wayne Dennis. Using as his source the National Academy of Sciences *Biographical Memoirs* for 1943–1952, he showed that of the 41 men who died after a full life, having reached the age of seventy, the top man had 768 publications, the bottom 27. The average number of publications was more than 200, and only 15 persons had fewer than 100 in their bibliographies. Similarly, a list of 25 eminent nineteenth-century scientists showed that all but one were in the range of 61 to 307 items.[4] Further, taking a sample from the *Royal Society Bibliography of Scientific Literature 1800–1900*, he showed that the most productive 10 percent of all authors, having each more than 50 publications, were of such caliber that 50 percent of them gained the distinction of mention in the *Encyclopaedia Britannica;* of the top 5 percent, each of whom had more than 140 items to his bibliography, some 70 percent received such mention. None of those mentioned in the *Encyclopaedia* by virtue of their scientific work had fewer than 7 publications.

Thus, although there is no guarantee that the small producer is a nonentity and the big producer a distinguished scientist, or even that the order of merit follows the order of productivity, there is a strong correlation,[5] and we are interested in looking deeper into the relative distribution of big- and small-output writers of scientific literature. Such studies are easy to make by counting the number of items under each author's name in the cumulative index of a journal. A pioneer investigation of this sort was made by Lotka,[6] and several others have since repeated such head counts. They all confirm a simple, basic result that does not seem to depend upon the type of science or the date of the index volume; the only requirement is that the index extend over a number of years sufficient to enable those who can produce more than a couple of papers to do so.

The result of this investigation is an inverse-square law of productivity (fig. 2.1). The number of people producing n papers is proportional to $1/n^2$. For every 100 authors who produce but a single paper in a certain period, there are 25 with two, 11 with three, and so on. Putting it a little differently by permitting the results to cumulate, one achieves an integration that gives approximately an inverse first-power law for the number of people who produce more than n papers; thus, about 1 in 5 authors produces 5 papers or more, and 1 in 10 produces at least 10 papers (fig. 2.2).

It is surprising that such a simple law should be followed so accurately and that one should find the same distribution of scientific productivity in the early volumes of the Royal Society as in data from the twentieth-century *Chemical Abstracts*. The regularity, I suggest, tells us something about the nature of the scores we are keeping. An inverse-square law probability distribution, or an inverse first power for the cumulative probability, is nothing like either the normal Gaussian or Poisson distributions, or any of the other such curves given by normal linear measure of events that go by chance. If the number of scientific papers were distributed in a manner similar to that of the number of men with various heights, or the number kicked to death by horses, we should find far fewer large scores. Scientific papers do not rain from heaven so that they are distributed by chance; on the contrary, up to a

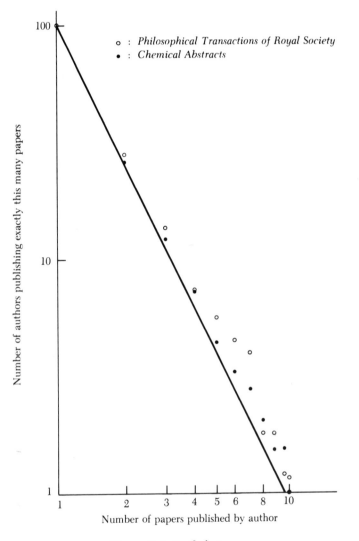

Figure 2.1. Lotka's Law

The number of authors publishing exactly n papers, as a function of n. The open circles represent data taken from the first index volume of the abridged *Philosophical Transactions of the Royal Society of London* (seventeenth and early eighteenth centuries), the filled circles those from the 1907-1916 decennial index of *Chemical Abstracts*. The straight line shows the exact inverse-square law of Lotka. All data are reduced to a basis of exactly 100 authors publishing but a single paper.

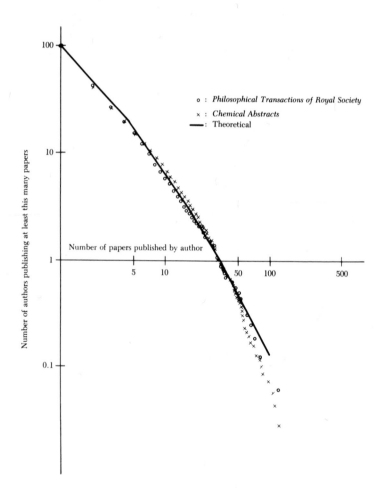

Figure 2.2. Number of Authors Publishing at Least *n* Papers as
a Function of *n*

Same data, and same reduction as for figure 2.1, but full curve here is modified
to a form that takes account of Lotka's overestimation of the number of highly
prolific authors (see note 8, chapter 2).

point, the more you have the easier it seems to be to get the next, a principle to which we shall return later.

Let us first examine the nature of the crude inverse-square law of productivity (table 2.2). If one computes the total production of those who write *n* papers, it emerges that the large number of low producers account for about as much of the total as the small number of large producers; in a simple schematic case, symmetry

Table 2.2
Schematic Table Showing Numbers of Authors of Various Degrees of Productivity (in papers per lifetime) and Numbers of Papers So Produced

Papers per Man	Men	Papers	
1	100	100	(The 75 percent of men who
2	25	50	are low scorers produce one-
3	11.1	33.3	quarter of all papers.)
4	6.2	25	
5	4	20	
6	2.8	16.7	
7	2	14.2	
8	1.5	12.5	
9	1.2	11.1	
10	1	10	
10–11.1	1	10+	
11.1–12.5	1	11.1+	
12.5–14.2	1	12.5+	
14.2–16.7	1	14.2+	(Subtotal: 10 men produce
16.7–20	1	16.7+	more than 50 percent of all
20–25	1	20+	papers)
25–33.3	1	25+	
33.3–50	1	33.3+	
50–100	1	50+	(The top two men produce
Over 100	1	100+	one-quarter of all papers.)
Total	165	586+	

Average papers per man = 586/165 = 3.54

Note: Table constructed on basis of exactly 100 men with a single published paper. Other entries computed from Lotka's law.

may be shown to a point corresponding to the square root of the total number of men, or the score of the highest producer. If there are 100 authors, and the most prolific has a score of 100 papers, half of all the papers will have been written by the 10 highest scorers, and the other half by those with fewer than 10 papers each. In fact, in this ideal case, a full quarter of the papers have been written by the top 2 men, and another quarter by those who publish only 1 or 2 items.

This immediately gives an objective method for separating the major from the minor contributors. One may set a limit and say that half the work is done by those with more than 10 papers to their credit, or that the number of high producers seems to be the same order of magnitude as the square root of the total number of authors. The first way, setting some quota of 10 or so papers, which may be termed "Deans' method," is familiar enough; the second way, suggesting that the number of men goes up as the square of the number of good ones, seems consistent with the previous findings that the number of scientists doubles every 10 years, but the number of noteworthy scientists only every 20 years.

Unfortunately, Lotka's simple inverse-square law needs modification in the case of high scorers (fig. 2.3). Beyond the divison lines mentioned, the number of people falls off more rapidly than the inverse square, more nearly approximating the inverse cube. It amounts to the same thing to say that their *cumulative* number falls off as the square of the score rather than as its first power. The data from the work of Lotka and Dennis agree completely on this, i.e., if one ranks the high scorers in order of merit, their scores fall as the square root of the ranks in all cases.[7]

By means of this one may easily derive a law which holds both for the low and high scorers and which slightly cuts down the upper tail of the Lotka distribution.[8] One can see that this should

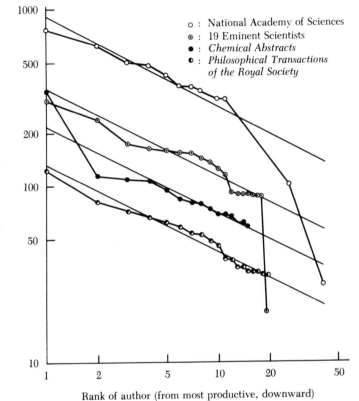

Figure 2.3. Numbers of Publications of Four Series of Highly Distinguished and (Incidentally) Highly Prolific Authors, Each Ranked Within the Series

The series are (1) members of the National Academy of Sciences, drawn from obituary bibliographies, (2) nineteen eminent scientists of the nineteenth century, (3) most prolific authors in decennial index of *Chemical Abstracts*, (4) index to vols. 1–70, *Philosophical Transactions of the Royal Society.*

be qualitatively necessary, since otherwise the maximum scores of published papers in a lifetime would be thousands and even tens of thousands rather than the several hundreds that seem to represent even the most prolific scientific lives. Cayley, one of the great British mathematicians of the nineteenth century, has 995 items in his collected works—a paper every two to three weeks—and I have failed to find anyone who outstrips this.

This modified law leads to the result that about one-third of the literature and less than one-tenth of the men are associated with high scores. It leads, furthermore, to an average of 3½ papers per man. Thus, if we know how many papers are published in a field, we can compute the number of men who have written them, even the much smaller number who must be reckoned as distinguished contributors to that field. Thus, for a field containing 1000 papers, there will be about 300 authors. About 180 of them will not get beyond their first paper, but another 30 will be above our cutoff of 10 papers each, and 10 will be highly prolific, major contributors.

More important than numerical information is the proved existence of a workable distribution law. One may make an interesting comparison between this and the famous Pareto law of distribution of income.[9] Instead of a form following $1/n$ for small values and $1/n^2$ for large, Pareto found that cumulative figures for income followed, almost exactly, and constantly over a long time in different countries, a law of $1/n^{1.5}$—just midway between our two forms. Why is there such an empirical law, and why is it so very different from the usual laws of errors, horse kicks, and other probability distributions?

The reason lies, I think, in the simple fact that the number of publications is not a linear additive measure of productivity in the way required for Gaussian distributions. Our cutoff point is not the average of the highest score and the lowest but rather their geometric mean. One feels intuitively that the step from 3 papers to 6 is similar to that from 30 to 60 rather than that from 30 to 33. Because of all this it is reasonable to suggest that we have here something like the approximate law of Fechner or Weber in experimental psychology, wherein the true measure of the response is taken not by the magnitude of the stimulus but by its logarithm; we must have equal intervals of effort corresponding to equal ratios of numbers of publications.[10]

We may define a man's solidness, s (how solid a fellow is he?) as the logarithm of his life's score of papers. The logarithm of the number of men having at least s units of solidness of productivity will at first fall linearly with s, then more rapidly as it approaches the fixed upper limit of 1000 papers, beyond which no man has achieved. In other words, for every unit increase in solidness, the number of men attaining such solidness is cut by some almost constant factor. Now this fall of the population by a constant factor for each unit increase of s is exactly what one finds in the tail of a normal probability distribution. For example, if we take the standard AGCT intelligence test distribution, which is so arranged that the norm is 100 on the scale, with half the population above and half below, and a spread such that the quartiles are at 80 and 120 (i.e., the standard deviation is 20), then for scores over about 140 (and also less than 60) the number of cases in the tail drops by a factor of 10 for every 10 points on the scale. If we measure solidness by logarithms to base 10, then every unit of s corresponds to about 11 points on the AGCT scale for all but the most solid scientific citizens, and for these it rises to about 20 points.

Pareto's law may therefore be regarded as merely the result of combining a reasonable probability distribution of capabilities with a Fechner's law measure of the effectiveness of these capabilities. In the case of scientific productivity we find a similar happy accounting on a theoretical basis for the shape of the empirical law. The only difference between the distributions of money and papers, or the more generalized distribution found by Zipf to account for nearly all natural distributions of things ranked in order of size, is that for science there is a definite upper limit to the amount that one man can accomplish in a lifetime.

Our one remaining uncertainty about the new law of normal distribution of scientific solidness is that we do not know where to put the beginning of the scale. What AGCT score corresponds to the state $s = O$, the minimal state of one scientific publication during a lifetime? If, without altering the presumably absolute and objective minimum standards for a scientific paper, one could induce every member of the population to go through the motions of education and professional training, and try to achieve this goal, how many would succeed?

This question is extraordinarily difficult to answer, for apart from a great corpus of general intelligence tests the competence level of the quantitative art is low when applied to deciding what makes for scientific creativity. On the basis of our newly won theory, one can now hazard a guess from intelligence tests alone. The fundamental investigations by Harmon on records of the United States crop of Ph.D.'s for 1958 enable us to say something of the incidence and of the intelligence test characteristics of this group.[11] Now, the Ph.D. and the editorial standards of learned periodical publications are things that we have done our best to keep constant. It is therefore reasonable to identify the minimum effort of writing a single scientific paper with that demanded by the "sheepskin gateway" to the road of research. Although it is agreed that these things do not coincide, since some Ph.D.'s never publish even their theses, whereas many authors are not doctors, yet at worst they should differ by some reasonably constant ratio not too far from unity.

Harmon found that in an age group of the population numbering

about 2,400,000 there arises an annual crop of about 8000 Ph. D.'s
in all fields, the physical and biological sciences together comprising about half the total. As one might expect, the intelligence
test scores for this group were considerably higher than the general level, the average being AGCT 130.8 for the mode of the
distribution. Taken by fields,, there was a variation from 140.3
for physics to 123.3 for Ph. D.'s in education:

Physics	140.3
Mathematics	138.2
Engineering	134.8
Geology	133.3
Arts and humanities	132.1
Social sciences	132.0
Natural sciences	131.7
Chemistry	131.5
Biology	126.1
Education	123.3

When these data were applied to the general population in the
same age group, it appeared that at the highest level of intelligence recorded, AGCT 170+, about 1 person in 5 received a
Ph. D., although the general incidence of doctorates in the age
group was only 1 in 3000. Thus, intelligence has a lot to do with
the gaining of Ph. D.'s. If we now consider it plausible that this
current figure of 1 in 5 refers to those superior beings who become
highly productive scientists, one could contemplate using all
means, fair and foul, to close the gap so that they would all earn
Ph. D.'s or even scientific Ph. D.'s.

We know now that the total number of scientists goes up as the
square, more or less, of the number of good ones. Therefore, if
we want to multiply the good scientists by five, we must multiply
the whole group by 25. Instead of an age group of about 8000
Ph. D.'s in mixed subjects, we should then have about 200,000,
all in science. As it happens, the intelligence distribution shows

that in an age group of 2,400,000, a few more than 160,000 achieve AGCT 130, and so we have a minimal cutoff for possible scientists that is only slightly less than the present mode found for Ph.D.'s, both scientific and otherwise. The two methods thus coincide to indicate that about 6 to 8 percent of the population at most could be minimal scientists.

Apparently, then, the scale of solidness in scientific publication should have its zero placed at an AGCT level of about 130, corresponding to about 1 person in every 15 in an age group. Attractive though it may be to perceive such a cutoff point, agreeing as it does so well with the present norm for Ph.D.'s, the implications are grave. At first sight it appears that at present we are tapping only about 1 in 25 of those who could become scientists at all, and a fifth of those who would be outstanding scientists. If we took all the talent of the population with no loss or wastage, we should then have 8,000,000 scientists writing papers in the United States,[12] and, of these, 80,000 would be highly productive, with more than 10 papers each. Thus, we should have a roll of 40,000 scientists to a million population, and, of these, 400 in a million would be men of note. Galton, you remember, found about 5 to 10 eminent scientists in a million population, and the early volumes of *American Men of Science* showed 50 in a million. Thus, in the density of good scientists we have left one more order of magnitude at the most and, even at the expense of all other high-talent occupations, science is not likely to engross more than 8 percent of the population. Even so, it looks as if the decreasing return of good scientists to every 100 Ph.D.'s will make it more and more difficult to reach a level of this magnitude. Just how strong is this limitation? Is it possible that the level of good scientists cannot rise by the factor of 5 that we have presumed?

Almost half of the factor is accounted for by the wastage of scientific womanpower, a wastage that the Soviet Union has partially checked but that we seem unable to avoid. Another factor of 2 might be attributed to the lack of opportunity and incentive in regions outside the big cities where schools are good and competition and inspiration keen. Indeed, all things considered, the high proportion of talented manpower successfully diverted into science at present is surely to our credit. But if the level cannot indeed rise, then we are, as we have already conjectured, about halfway toward saturation at the top end of the scale, and any increase in numbers of scientists must produce an even greater preponderance of manpower able to write scientific papers, but not able to write distinguished ones. It gives serious pause to thoughts about the future of scientific education. Is it worth much sacrifice?

I think we have now laid the theoretical basis for this study of science. It is remarkably similar to the study of econometrics. On the one hand, we have the dynamic treatment that gives us time series, first of exponential growth, then of the saturated growth resulting in standard logistic curves. On the other hand, we have the statics of a distribution law similar to that of Pareto. The extent of the difference between analyzing science and analyzing business lies in the parameters. The main exponential part of the growth of science doubles in 10 years only, which is much more rapid than all else; the characteristic index of the distribution law is 1 at the low end and 2 at the high, instead of a uniform 1.5.

The additional contributions that we have made lie in providing a reasonable theoretical basis for our Pareto law and in showing that, although the average number of papers per author remains sensibly constant, one may make a split between those whose productivity is high and that much larger mass of authors whose productivity is low. This mass is seen to grow as the square of the number of high scorers, and therefore the number of high scorers will appear to double only every 20 years.

The Fechner law principle which we invoked to reduce the Pareto-like distribution to the sort of linear and additive measure that is necessary for a standard probability curve is much more powerful than we have yet assumed. If we may take in general the solidness of a body of publications as measured by the logarithm of the number of papers, it has further interesting consequences. Consider the law of exponential growth previously mentioned as a universal condition of freely expanding science. Obviously, the solidness of the field, the logarithm of the number of papers, grows linearly with time. Thus, since it takes about 50 years for the number of men or number of papers in a field to multiply by 10, there is a unit increase of solidness every half-century.[13]

I cannot quite see why it is so, or how one might judge it other than by pure intuition, but the two units of solidness separating the man who can publish no more than one paper in a lifetime from the one who can write a hundred such papers are essentially the same as those that separate the two states of a subject at dates a century apart. In rough, and misleading, terms one might say that the eminent scientist is a century ahead of the minimal one.

What further implications are there of the assumption that one can measure the progress of a field by the linear march of its solidness? Are such degrees of solidness truly additive? Must we judge one field of a hundred workers adding two units of solidness within a certain time as inferior to ten separate fields of ten workers, each of whom will add one unit to each field, making a total of ten units within the same time?

If such indication be true, then it seems that science has a strong desire to minimize its solidness rather than make it as large as possible. Beyond the phenomenon of exponential growth, science displays in several ways a tendency to crystallize out, in the sense that big things grow at the expense of the small ones that constitute a sort of mother liquor. Large fields seem to absorb the manpower and subject matter of small ones. Even though new fields, new departments, new institutions, and even new countries arrive on the scientific scene in increasing number, the few previously existing large ones have a natural growth enabling them in general to maintain their lead. It is the exception, rather than the rule, for one of the big blocks to slacken its growth—presumably through the existence of some sort of logistic ceiling that causes it to stagnate—and be overtaken so that it falls in rank.

The fact that the general growth of science increases equally the sizes of the large blocks and the numbers of the small blocks, while presenting an appearance of crystallization, is really not so peculiar. Precisely the same thing happens when the population of a country grows. Instead of being uniformly distributed over the country, it is crystallized out into variously sized blocks called cities. The growth of cities in a country provides a useful model for the growths of scientific blocks within science. As it happens, the hierarchical order of cities or other blocks, ranked by decreasing size, offers yet another example of the same Pareto-like distribution we have already found for the productivity of scientific authors.

In the case of cities, the historical statistics provide a good example of such a distribution on the move, with everything increasing exponentially while maintaining the normal distribution (fig. 2.4).[14] Using a plot showing the distribution at each decade, one may see the constant slope of the distribution on a log-log scale and the inexorable march of the intercepts that tell us the magnitude of the biggest city on the one scale and the number of minimal cities (here taken as population 2500) on the other scale. Both increase regularly each decade, taking about sixty years each to go through a power of 10 or, as we have called

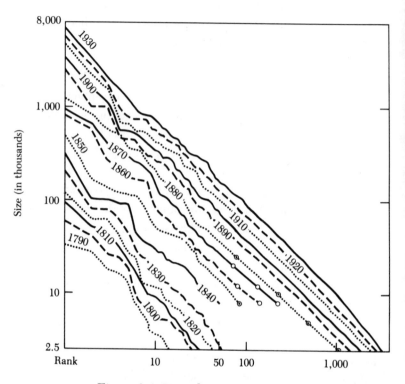

Figure 2.4. United States, 1790–1930

Communities of 2500 or more inhabitants ranked in the decreasing order of population size. It should be noted that the distribution at any given date shows size decreasing uniformly with rank; as cities become more numerous and all of them increase in size, the distribution pattern is preserved, the curve moving parallel to itself at a constant rate. From George K. Zipf, *Human Behavior and the Principle of Least Effort* (Cambridge, Mass.: Addison-Wesley, 1949), p. 420, figure 10-2.

it before, one unit of solidness. If one looked in detail at the life history of any particular city, its rank would change with time as it outpaced others and was itself outpaced; however, the statistical distribution is remarkably constant.

This general pattern, carrying all the implications of our previous analysis of productivity distribution, is followed fairly well by such diverse hierarchical lists as those giving the sizes in faculties, or in Ph.D.'s per decade, of the college scientific departments, in any field or in general, in the United States or in the world. It is followed by ranked lists showing the scientific contributions, in terms of papers, journals, or expenditures of the nations of the world, ranging from the few big producers on any scale relative or absolute to the minor production of the large number of underdeveloped countries (fig. 2.5).[15]

About this process there is the same sort of essential, built-in *undemocracy* that gives us a nation of cities rather than a country steadily approximating a state of uniform population density. Scientists tend to congregate in fields, in institutions, in countries, and in the use of certain journals. They do not spread out uniformly, however desirable that may or may not be. In particular, the growth is such as to keep relatively constant the balance between the few giants and the mass of pygmies. The number of giants grows so much more slowly than the entire population that there must be more and more pygmies per giant, deploring their own lack of stature and wondering why it is that neither man nor nature pushes us toward egalitarian uniformity.

Value judgments aside, it seems clear that the existence of a reasonable distribution that tells us how many men, papers, countries, or journals there are in each rank of productivity, utility, or whatever you will measure provides a powerful tool. Instead of attempting to get precision in defining which heads to count in

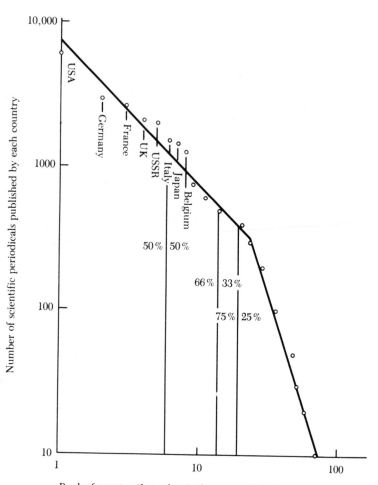

Figure 2.5. Number of Scientific Periodicals Published by Each
Country, Ranked in Decreasing Order of Such Numbers

It will be seen that the top six countries account for half of all publications, the
top eleven for two-thirds, etc. Productivity of journals falls off very rapidly for the
less prolific countries. Data from a preliminary survey conducted by the Library
of Congress.

exponential growth, one may instead take a crude count and interpret it by means of such a distribution.

Just as one cannot measure the individual velocities of all molecules in a gas, one cannot actually measure the degrees of eminence of all scientists. However, there are reasonable grounds for saying that such measurements, if made, would follow the standard distribution. In particular, we can take this Pareto-like distribution as a hypothesis and see how the consequences agree with the gross phenomena which we can measure. We do, in fact, find a reassuring agreement.

Such, then, is the broad mathematical matrix of exponential growth, logistic decay, and distribution functions. It provides us now with a general description of the normal expansion of science and its state at any time. Knowing now the regular behavior, we have a powerful tool for investigating the significant irregularities injected into the system by the gross perturbations of war and revolution, by the logistic birth and death of measurable entities, by genius and crucial discovery, and, in short, by the organizational changes within the body politic of science and in its relations with the state and society in general.

3

Invisible Colleges and the Affluent Scientific Commuter

From all the talk of exponential growth and scientific productivity distributions, one might think that scientific papers were produced merely to be counted by deans, administrators, and historians and that the driving force of a scientist should be directed toward producing the maximum number of contributions. This is far from the truth. An almost instinctive reaction away from all this counting nonsense is to agree that each paper represents at least a quantum of useful scientific information and that some single contributions may rise so far above this quantum value that for such a one alone its author would be valued above any random hundred, or even above a hundred more prolific writers.

To take the opposite point of view and look into the tangible results of scientific work more deeply than by mere head counting, we must know considerably more about the social institutions of science and the psychology of the scientist. The prime object of the scientist is not, after all, the publication of scientific papers. Further, the paper is not for him purely and simply a means of communicating knowledge.

Let us look at the history of the scientific paper. It all began because there were too many books. Here is cry from the heart of a scholar:

One of the diseases of this age is the multiplicity of books; they doth so overcharge the world that it is not able to digest the abundance of idle matter that is every day hatched and brought forth into the world.

It is chastening to find that these words were written by the rambunctious Barnaby Rich in 1613, half a century before the scientific journal was born. The coming of the learned periodical promised an end to this iniquity of overcharge. Developing in time and spirit together with the newspaper, such publications as the *Philosophical Transactions of the Royal Society* had the stated function of digesting the books and doings of the learned all over Europe. Through them the casual reader might inform himself without the network of personal correspondence, private rumor, and browsing in Europe's bookstores, formerly essential.

At first, however, they did not by any means remit the scholarly obligation to read books and write them. Their original purpose was a social one of finding out what was being done and by whom rather than a scholarly one of publishing new knowledge.[1]

Original publication of short papers by single authors was a distinct innovation in the life of science and, like all innovations, it met with considerable resistance from scientists. Barber has pointed out that such resistance is part of a vital mechanism of innate conservatism in the body of science.[2] It is a natural counterpart to the open-minded creativity that floods it with too many new ideas, and to the edge of objectivity that forms an eventual means of deciding between true and false.

Such resistance against the new and seemingly illicit practice of publishing papers instead of decent books is seen in the case of Newton. The controversies over his optical papers in the *Philosophical Transactions* were a source of deep distress to him, and afterward he did not relish publication until it could take the proper form of a finished book, treating the subject from beginning to end and meeting all conceivable objections and side arguments. If the journal had been at that time an effective means of communication, we might never have had the *Principia*. Perhaps we should begin to disregard a man's papers and look at his books.

The transformation of the scientific paper into its modern state was not complete until about a century ago. Before that time there was much publication of scientific snippets, such as the bare mention of something achieved, or a review of observations that had been made and published elsewhere. There were also plenty of monographic publications that would have been books in themselves if only the means for profitable printing and distribution had existed. As late as 1900, some of the most respected journals contained not one scientific paper of the present variety. The difference is not only one of length—if they are too short, they are letters; if too long , monographs.

I would rather make a distinction in the mode of cumulation of the papers. This has to do with the way in which each paper is built on a foundation of previous papers, then in turn is one of several points of departure for the next. The most obvious manifestation of this scholarly bricklaying is the citation of references. One cannot assume that all authors have been accurate, consistent, and conscientious in noting their sources. Some have done too little, others too much. But it is generally evident from a long run of any scientific periodical that around 1850 there appears the familiar modern pattern of explicit reference to previous work on which rests the distinct, well-knit addition that is the ideal burden of each paper. Before that time, though footnoting is as old as scholarship itself—compare the very term *scholia* for the ancient footnote—there is nothing like this attitude toward the accretion of learning.

If, then, the prototype of the modern scientific paper is a social device rather than a technique for cumulating quanta of information, what strong force called it into being and kept it alive? Beyond a doubt, the motive was the establishment and maintenance of intellectual property. It was the need which scientists felt to lay claim to newly won knowledge as their own, the never-gentle art of establishing priority claims.

In a pair of perceptive papers, Robert Merton has analyzed the way in which priority claims and disputes have been omnipresent during the past few centuries of science.[3] The phenomenon emerges as a dominant thread in the history of science, woven through the stories of all men in all lands. It is fair to say that to understand the sociological character of such disputes is more important for the historian than merely to settle such claims.

The evidence makes it plain that multiple discovery—that is, discovery by two or more individuals working separately—occurs with remarkable frequency, that it often gives rise to disputes for priority among the parties concerned, and that these disputes may be laced with the bitterest and most violent passions of which the protagonists are capable. Several important things about the life of science may be learned from this analysis.

First, the multiplicity of discovery runs so high in so many cases that one is almost persuaded that it is a widespread occurrence rather than a chance rarity. As Galton remarks, "When apples are ripe they fall readily." One may go further, as Kuhn has done,[4] and remark that although some discoveries, such as X rays or oxygen, take one completely by surprise, there are many, many more which are more or less expected, and toward which several people are working simultaneously. It is in the latter class that we experience the multiple discovery and the disputed priority, though probably the disputants would hotly contest that their prize discovery was in no wise expected and that their opponent had either stolen the idea or had discovered only inadvertently a part of the essential new matter.

The figures cited by Merton and Barber for the historical incidence of multiple discovery in various degrees enable us to test, in a fashion highly instructive, the "ripe apple" model.[5] If there are 1000 apples on a tree, and 1000 blindfolded men reach up at random to pick an apple, what is the chance of a man's getting one to himself, or finding himself grasping as well the hand of another picker, or even more than one? This is a straightforward question in statistical probability. By means of the Poisson distribution it is found that 368 men will be successful and that 264 cases will involve the remaining 632 men in contested claims (table 3.1).

The agreement betwen expectation and fact, at least for the doublet, triplet, and quadruplet discoveries, is striking but must not be given too much credence. To fit the data we have made two arbitrary assumptions: first, that we start with 1000 pickers or discoverers; second, that there shall be on the average one prize for each. The first assumption is reasonable, for one cannot avoid adjusting the data to some sort of total population. The second is harder to justify, especially as it involves 368 apples that were not picked at all, discoveries that were missed because of the overlapping hands. As a first approximaton, however, we note that only 37 percent of the seekers will establish uncontested claims, the remaining 63 percent will end in multiple discovery. In terms of actual discoveries made, the positon is a little brighter:

Table 3.1
Poisson Distribution and Simultaneous Discovery

Number of Simultaneous Discoverers	Merton Data Cases	1000 Apples and Men Cases
0	Indeterminate	368
1	No data	368
2	179	184
3	51	61
4	17	15
5	6	3
6 or more	8	1

about 58 percent will be unique, and only 42 percent will be shared by two or more.

As a second approximation, the data show more instances than we expect by random choice involving five or more coincident pickers of the same discovery. Perhaps the apples that appear to be biggest and most ready to drop attract more than their due share of the pickers, but this is only a minor amendment to the gross phenomenon.

Not all cases of multiple discovery end in hotly contested priority disputes. Merton shows that the tendency has decreased as we have become used to the idea that this is bound to happen, the proportion of disputes being 92 percent in the seventeenth century, 72 percent in the eighteenth, 59 percent by the latter half of the nineteenth, and 33 percent in the first half of the present century.

Even at these rates, the passion generated and the large amount of overlapping discovery that seems to have been with us throughout the recorded history of the scientific paper makes us wary of the role of that device. If it is for frontline communication, then we must feel that it has always done a remarkably poor job of preventing overlapping researches. The applepickers appear to act as if they were blindfolded to the efforts of others rather than as if they had any information in time for them to move their hands to one of the many untouched fruits. If paper publication is not for frontline communication, let us cease to complain about overlapping.

The scientific paper therefore seems to arise out of the claim-staking brought on by so much overlapping endeavor. The social origin is the desire of each man to record his claim and reserve it to himself. Only incidentally does the paper serve as a carrier of information, an annoucement of new knowledge promulgated for the good of the world, a giving of free advantage to all one's competition. Indeed, in past centuries it was not uncommon for a Galileo, Hooke, or Kepler to announce his discovery as a cryptogram of jumbled letters that reserved priority without conferring the information that would help his rivals. In the present day, as Reif has pointed out, the intense competition to publish "fustest and mostest" and thereby achieve prestige has resulted in a long series of abuses and high emotions ranging from illicit publication in the *New York Times* to rare cases of fraudulent claim.[6]

Why the scientist acts in this way is another question. The answer to it, I feel, may involve some rather deep psychological analysis of the scientific character. At the root of the matter is the basic difference that exists between creative effort in the sciences and in the arts.[7] If Michelangelo or Beethoven had not existed, their works would have been replaced by quite different contributions. If Copernicus or Fermi had never existed, essentially the same contributions would have had to come from other people. There is, in fact, only one world to discover, and as each morsel of perception is achieved, the discoverer must be honored or forgotten. The artist's creation is intensely personal, whereas that of the scientist needs recognition by his peers. The ivory tower of the artist can be a one-man cell; that of the scientist must contain many apartments so that he may be housed among his peers.

Two important implications emerge from this analysis. First, scientific communication by way of the published paper is and always has been a means of settling priority conflicts by claim-staking rather than avoiding them by giving information. Second, claims to scientific property are vital to the makeup of the scientist and his institutions. For these reasons scientists have a strong urge to write papers but only a relatively mild one to read them. For these reasons there is a considerable social organization of scientists whose aim is to establish and secure the prestige and priority they desire by means more efficient than the traditional device of journal publication.

When one talks about the information problem in science, it is, I feel, important not to confuse the matter with that which we have just described. For three centuries science has lived effectively with the high incidence of multiple discovery and disputed claims for priority. At every turn in past history it was to be regretted that X's ideas were not known to Y. The overlapping could hardly have been worse, and there is no clear evidence that it has ever either improved or deteriorated.

Perhaps it is not just the counsel of despair to posit that science has lived vigorously if not happily on its diet of disputes and duplications. Perhaps it is even desirable that many of the important discoveries should be made two or three times over in an independent and slightly different fashion. Perhaps men must themselves recreate such discoveries before they can usefully and effectively go on to the next stage. We seem nowadays to dispute less about the same amount of overlapping, but perhaps we have only turned our wrath against the societies, publishers, librarians, and editors who seem to conspire to leave us in such a duplication-prone position. However, let us be fair. We may complain that they have not removed this stumbling block from our path, but we cannot well complain that it has grown worse. It could hardly be worse. Our information problem, assuming we have one, is of a different nature.

Let us first look at the organization problem of scientific literature in terms of the input and output of any one man. We have seen that the normal scientist may during his lifetime publish papers ranging in number from a minimal one up to several hundred and that the borderline between many and few is about the geometric mean between these limits. Consider now how much he must read in order to produce those papers. At the beginning of his career, his teachers and his basic reading of books and current literature in a chosen subject will have placed him at the research front, and from there he will perhaps be able to voyage alone on uncharted seas. If this man remains in a field of which he is the sole exponent, he can read nothing besides his own papers. Such is the life of the lone pioneer who has no need to read journals and publishes (if he does) only for the good of future generations.

But life is usually not like that. The man arriving at the research front finds others with the same basic training in the same subject looking at the same problems and trying to pick apples off the same tree. He will want to monitor the work of these similar individuals who are his rivals and his peers. He will want to leapfrog over their advances rather than duplicate them. How many such individuals can be so handled? I suggest that the answer is on the order of a hundred. Surely he can read one paper for every one he writes. Just as surely, he cannot efficiently monitor 10,000 papers for each one of his own, at which rate the good man who writes 100 papers in a lifetime would be reading a million, or more than 60 a day.

Another way of deriving this ratio is to think of the number of people with whom a good scientist can exchange offprints, pre-prints, and professional correspondence, and with whom he can perhaps collaborate at a reasonable and comprehensive level. Publishers have their records of the purchase of reprints, but I know of no published figures. My guess is that there are a few hundred colleagues for every worker. Here, of course, we are dealing with numbers of *actual* men rather than with numbers of papers due to *effective* men. We do read several papers by people who are not on our lists, after all, and correspondingly ignore some of the output by our friends.

There is yet another way of looking at this ratio. The norm of the number of papers given as references in a research paper has for many years been constant at a little less than 10. Supposing we read, closely enough to cite them, about 10 papers for every one we actually cite, there would then have been about 100 papers read for every one published. Our tendency to faithfully repeat citations of our favorite and most useful papers only reduces this figure.

It seems, then, that we can handle an effective input that is little more than a few hundred times the size of our output. Perhaps those who write little have more time for reading than those who are prolific, so that there is some sort of balance. Perhaps the true research man does not read at all but takes his input in other ways, orally and socially. On the whole, one can keep up with a colleague group that has an effective size of a few hundred members; one cannot possibly keep up with 10,000.[8]

However, since all aspects of science grow exponentially with the remarkable rapidity of a factor of ten in fifty years, it seems clear that when a subject has reached the stage where its first dozen protagonists are beginning to feed on one another's papers and to watch their priorities and advances, it can scarcely be expected to remain intact as a field for another generation. When in the course of natural growth it begins sensibly to exceed the few hundred members postulated, each man will find himself unable to monitor the field properly.

At each stage along the way the backlog of papers can be packed down into review articles and eventually into textbooks. For example, the progenitor of such a field, looking back at the end of his working life upon, say, 100 papers of his own and an effective list equivalent to 100 colleagues, can muster a bibliography of 10,000 items, duly compressed into a critical review of the state of the art. But this never solves the current problem of more than several hundred men trying to keep up with one another's work.

One of the traditional modes of expression among such groups is the founding of a new scientific organ, a journal which is their medium for communication. A membership of several hundred may be augmented by a thousand or more individuals only fractionally or marginally within the group. Add to this the subscription list of the libraries which decide that the journal is necessary to them and the usual quota of miscellaneous subventions, and one has an economic modicum for such a publishing endeavor.

This gives us, incidentally, a check on our ratio of 100. Since science began, about 10 million scientific papers have been published, and we are adding to them, with a doubling in 10 years, or about 6 percent a year, about 600,000 new papers every year. These come out in some 30,000 current journals, which therefore carry an average of 20 per year. Now 10 million papers implies the existence of about 3 million authors, most of whom, because of exponential growth, are alive now. Therefore, there is approximately 1 journal for every 100 authors. Since the seventeenth century, the besetting sin of all journal creators has been to imagine that theirs was a journal to end all journals in that particular realm of subject matter.[9] One doubts whether any group like the audience of such a journal has remained a closed set beyond the appearance of the first issue. Members of the group invariably read more papers than those prescribed for them by their colleague editor. Moreover, members of other groups find that their diet can be improved by reciprocal poaching. Thus, although there is an average of only 100 scientists to each journal, nevertheless, it will reach about 1000 scientists if each man looks at 10 serials.

Such overlapping, as in multiple discovery, generates heat and lowers efficiency. What is sought is the adiabatic expansion that could be had if science could be compartmented into watertight areas, that is, if a man in one area need never extend his research reading into any other. But evidently science abhors such splitting. Even the splitting of chemistry from physics when the cake of natural philosophy was divided gave rise automatically to disciplines of physical chemistry and chemical physics, so that each section needed constant surveillance of the others adjacent. Overlap of research fields is a sort of embargo that nature exerts against the urge that man has to divide and conquer.

As might be expected, journals are not shared with 10 men reading each issue or each paper. A now classical paper by Urquhart analyzed the crop during 1956 of 53,000 external loan requests filled by the (central) Science Library in London from its holdings of 9120 different scientific periodicals, of which more than 1300 were not current (fig. 3.1).[10] More than 4800 of the current titles were not used at all during the year; 2274 were used only once. At the other end of the scale, the most popular journal had 382 requests, 60 titles were requested more than 100 times each, and half the requests could be met from the top 40 journals. Less than 10 percent of the available serials were sufficient to meet 80 percent of the demand.

This distribution in rank of journals is equivalent to that which we have already met in scientific productivity. There is the same Pareto curve as in the distributions of incomes or sizes of cities, apparently for much the same reasons. Thus, journal-dwellers are distributed in the same way as city-dwellers; there is the same tendency to crystallize, and the same balance between the exponential growth of the largest members and the increasing numbers of the smallest. Since the dividing line is drawn at the square root of the total population, we can say that although 30,000 journals exist, half the reading that is done uses only the 170 most popular items.

Amount of use seems intuitively to be a better test of quality than our former criterion, amount of productivity. Unfortunately, though we now have figures for the utility of journals in terms of

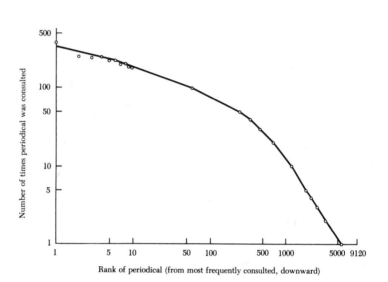

Figure 3.1. Utilization of Journals (i.e., Number of Times Consulted in a Given Year), Ranked in Decreasing Order of Utilization

In this study, more than 3000 of the available total of 9120 journals in the library were not called for at all during the period of investigation.

their rate of usage by a large population, we have no comparable figures for individual papers. It seems almost inevitable on qualitative grounds alone that the same conditions would apply, and that there would be a Pareto-like distribution linking a hierarchy of most popular papers at the top end of the scale with a low-ranking group used twice, or once, or perhaps never.

From this it would follow that all statements hitherto made

about the numbers of good researchers vis-à-vis poor researchers would apply if we had data for a true count of quality rather than the admittedly crude count of quantity. We know that the ranks of individuals would not by any means correspond on the two scales, but one could say with some assurance that there would be a significant correlation between qualitative solidness and quantitative solidness. However, since it is fortunately not incumbent upon us to provide such measures for individuals, all we need is the knowledge that the statistical mechanics of scientific manpower and literature obey such general laws.[11]

It also follows from that existence of these stable and regular distribution curves that we may now justify on a theoretical basis our previously empirical procedure of using crude numbers of periodicals or papers as an index of the size of science. We know now that any measure of total number of journals, papers, or men will give the corresponding number of important journals, papers, or people. It will be seen that a slight change in definiton—for example, uncertainty about the minimum allowable level at which a journal may be accounted scientific—will only increase the size of the tail. This is why even the loosest definitions yield usable results and regular exponential growths.

Having posited that amount of usage provides a reasonable measure of the scientific importance of a journal or a man's work, let us apply this to the scientific paper in general. Let us consider the use of a paper in terms of the references made to it in other papers. We shall have to ignore the evident malpractice of some authors in preferentially citing their own papers, those of their special friends, or those of powerful or important scientists that confer status on their work. We shall also take a rosy view in supposing that the practice of first writing the paper and then adding for decoration some canonical quota of a dozen references—like Greek pillars on a Washington, D.C. building—does not sensibly pervert the average conscientiousness in giving credit to papers that have provided the foundation for the work.

We suppose, then, that a research contribution is built from a man's own work, from a corpus of common knowledge needing no specific citation, and from an average of 10 other papers to which reference is made. Take now a field in which since the beginning of time a total of N papers has been published. If that field is doubling every decade, as healthy fields do, the next year will produce an additonal crop of $0.07 N$ papers, and these will contain $0.7 N$ references to the backlog of N papers. On an average, then, each of the N papers will be cited by new ones at the rates of 0.7 times per year. We have supposed, however, that the incidence of citation and referencing, since these measure the utility of the various papers, cannot be spread out uniformly. Some papers will be cited much more than others. Some may fall unnoticed and never be cited.[12]

Let us look first at the way in which citation appears to fall off with age. It has been remarked several times that if all the references cited in a single issue of a journal, or the volume for a certain year, are sorted according to date, then the number falls off rapidly as one goes back in time.[13] Fussler investigated physics and chemistry journals of various dates and showed that although papers as old as 150 years had been cited, there was clearly a falling-off with age.[14] Half of all references in chemistry were made to papers less than 8 years old, half in physics to papers published in the preceding 5 years. Unfortunately, the data are badly upset by his use of the war years 1919 and 1946 as half of the sample.

A better analysis of the useful half-life of papers can be made from the librarians' investigations of the amount of use given the various bound volumes of their runs of periodicals (fig. 3.2). In the greater libraries, among large populations of such journals, it

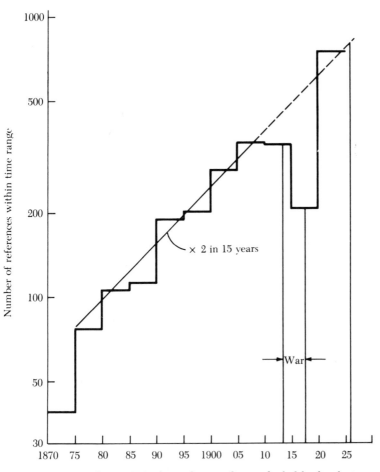

Figure 3.2. Count from a 1926 volume of a scientific periodical of the distribution by date of all references cited in that volume. It will be noted that with the exception of a five-year period embracing World War I, the number of references falls off by a factor of two in fifteen years.

has been found several times that the use falls off by a factor of two in times on the order of nine years. The data of Gross and Gross on references made in a single volume (1926) of *Chemical Literature,* show a halving of the number for every 15 years of increased age.[15]

Although this falling away is striking, remember that the actual amount of literature in each field is growing exponentially and therefore doubling every 10 to 15 years. Therefore, to a first approximation, the number of references of a given date seem to rest in proportion to the total literature available at that date. Thus, although half the literature cited will in general be less than a decade old, it is clear that, roughly speaking, any paper once it is published will have a constant chance of being used at all subsequent dates.[16]

This rather surprising result may be modified in improving our approximation. In fields tending to honor their pioneers by eponymic fame—name laws, name constants, name species—one may find that good papers actually improve with age, and their chance of citation increases. In fields embarrassed by an inundation of literature there will be a tendency to bury as much of the past as possible and to cite older papers less often than is their statistical due. This tendency can be seen in the journal *Physical Review Letters,* which achieves the greatest possible rapidity of publication.

In these *Letters,* since their foundation, the half-life of references has been stable at about 2½ years; that is to say, half of all references are younger than that. Now, the past 2½ years of physics literature contains less than one-third of all the work published during the last decade, and that decade, of course, contains half of all that has ever been printed. The people publishing in these letters are thus enabled by rapid publication to deal with less than one-third of all the papers that would normally be involved. To balance this, papers in the field must necessarily be cited something like three times as frequently, and therefore the amount of overlapping of citations is much increased.

Now, papers which make the same citations have an increased likelihood of doing the same work. Thus, increasing the efficiency of the rate at which one can make priority claims automatically seems to produce a higher incidence of such claims, or at least of the raw material for them. There is a feedback working to minimize part of the advantage gained by rapid publication.[17]

Let us look next at the way that references and citations are distributed other than in terms of date. If we were to rank any population of papers in terms of a hierarchy having at the top the most-cited paper of the year and at the bottom those given as a reference only once or not at all, we should evidently have a Pareto-type distribution similar to that found for the utility of scientific journals. In terms of this we might, if we had the information, say that half of all citations were given to a small group of papers existing at that time. On purely qualitative grounds one would suppose that 100 papers out of a field of 10,000 supply about one-third of the citations. On the other hand, there will inevitably be several thousand papers that are lost, or cited so rarely that they do not become generally known. It is impossible to say how much of this loss is deserved and just, but a large body of jilted authors will feel that it is not. There are cautionary tales of rediscovered papers, like that of Mendel, to make us feel that the statistical loss of literature must be minimized.

Thus, the essential problems of scientific manpower and literature are twofold. At the top the critical problem is predominantly one of human engineering: arranging for the highest level people to interact in manageable numbers, seeing that the great journals continue to correspond to large natural groups, arranging for the important papers to be collected and compressed into standard monographs and texts. At the lower end it is one of switchboard operation: how does one manage the large body of average scientists and appliers so that it keeps pace with the leaders; how does one monitor the lesser journals and the almost unnoticed papers so as to prevent wastage? We shall see several different mechanisms at work, each of them presently in a critical condition, as we make the logistic transition from Little Science to Big Science.

The first noteworthy phenomenon of human engineering is that new groups of scientists emerge, groups composed of our maximal 100 colleagues. In the beginning, when no more than this number existed in a country, they could compose themselves as the Royal Society or the American Philosophical Society. At a later stage, they could split into specialist societies of this size. Now, even the smallest branches of subject matter tend to exceed such membership, and the major groups contain tens and hundreds of thousands. In a group of such size, by our previous analysis, there are likely to be a few groups of magnitude 100, each containing a set of interacting leaders. We see now such groups emerging, somewhat bashfully, as separate entities.

Probably during World War II, pressure of circumstances forced us to form such knots of men and keep them locked away in interacting seclusion. We gave them a foretaste of urgent collaboration in nuclear physics, and again in radar. These groups are still with us in the few hundred people who meet in the "Rochester Conference" for fundamental particles studies, and in the similar number who congregate by invitation to discuss various aspects of solid state physics.

The organization is not perfect; a few of the best men may not attend, a few of those who do attend might not qualify if we had perfect objective judgment. Conscientiously, one might try not to be too exclusive, not to bar the gentleman from Baffinland who would be a distinguished researcher on fundamental particles if only he could. But there is a limit to the useful size, and if too many are invited, an unofficial subgroup of really knowledgeable members will be forced into being.

Such activity is by no means confined to the two groups mentioned. Similar unofficial organizations exist in molecular biology, in computer theory, in radio astronomy, and doubtless in all sciences with tens of thousands of participants. By our theory they are inevitable, and not just a product of the war or the special character of each discipline. Conferences are just one symptom; it becomes insufficient to meet as a body every year, and there is a need for a more continuous means of close contact with the group of a hundred.

And so these groups devise mechanisms for day-to-day communication. There is an elaborate apparatus for sending out not merely reprints of publications but preprints and pre-preprints of work in progress and results about to be achieved.[18] The existence of such a group might be diagnosed by checking the preprint list of one man and following this by a check of the list of each man mentioned. I think one would soon find a closed group, a small number of hundreds in membership strength, selected from a population of a large number of tens of thousands.

In addition to the mailing of preprints, ways and means are being found for physical juxtaposition of the members. They seem to have mastered the art of attracting invitations from centers where they can work along with several members of the group for a short time. This done, they move on to the next center and other members. Then they return to home base, but always their allegiance is to the group rather than to the institution which supports them, unless it happens to be a station on such a circuit. For each group there exists a sort of commuting circuit of institutions, research centers, and summer schools giving them an opportunity to meet piecemeal, so that over an interval of a few years everybody who is anybody has worked with everybody else in the same category.

Such groups constitute an invisible college, in the same sense as did those first unofficial pioneers who later banded together to found the Royal Society in 1660. In exactly the same way, they give each man status in the form of approbation from his peers, they confer prestige, and, above all, they effectively solve a communication crisis by reducing a large group to a small select one of the maximum size that can be handled by interpersonal relationships. Such groups are to be encouraged, for they give status payoff without increasing the papers that would otherwise be written to this end. I think one must admit that high-grade scientific commuting has become an important channel of communication, and that we must ease its progress.

Possibly, if such groups were made legitimate, recognized, and given newspaperlike broadsheet journals circulating to a few hundred individuals, this would spoil them, make them objects of envy or of high-handed administration and formality. Elite scientific newspapers or broadsheets of this sort have long existed in Japan, a country faced with the special problem that many of its top scientists spend appreciable periods in foregin institutes.

The scientific elite have acquired prestige among the public in general and the employers in particular, which has given them a certain affluence and enabled them to commute. It incidentally replaces the kudos they have lost since the debasement of the coinage of scientific publication. Despite a tendency to place summer schools in pleasant resort areas whenever possible and to make institute housing a good place to bring one's family, there is a further need. There is a further need to recognize that although a place such as Brookhaven was once where one went to work with big machines and certain other facilities, it has come nowadays to play an increasingly important role as a station on the commuting circuit of several invisible colleges. People come to work with other people, who have come to work with yet other people, who happen to be there. We need many more such facilities in various fields and in various countries. It might, for example, be wise for the United States government to subsidize the erection of "Fulbright residential buildings" in London, Cambridge and Oxford, Copenhagen, Geneva, Paris, Delhi, and wherever else United States scientists habitually commute in quantity.

So much for the elite—what of the masses? Mention of the big machines is immediately reminiscent of one way in which the formation of elites is producing a problem in the organization of the rest of the scientific population. It has become common to organize research, especially big machine work, around quite a large team of men comprising a few leaders in various specialties and a large number of younger men. Now it becomes the custom to publish as just such a team. As an editor of *Physical Review Letters* plaintively noted on a recent occasion, "The participating physicists are not mentioned, not even in a footnote."[19]

Surprisingly enough, a detailed examination of the incidence of collaborative work in science shows that this is a phenomenon which has been increasing steadily and ever more rapidly since the beginning of the century (fig. 3.3). It is hard to find any recent acceleration of the curves that would correspond to the coming of the big machine and indicate this as a recognizable contributing cause.

Data from *Chemical Abstracts* show that in 1900 more than 80 percent of all papers had a single author,[20] and almost all the rest were pairs, the greater number being those signed by a professor and his graduate student, though a few are of the type Pierre and Marie Curie, Cockcroft and Walton, Sherlock Holmes and Dr. Watson.[21] Since that time the proportion of multiauthor papers

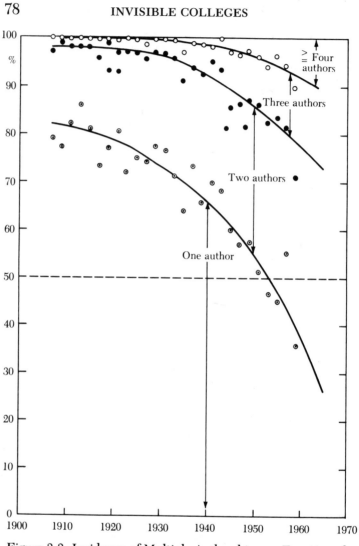

Figure 3.3. Incidence of Multiple Authorship as a Function of Date

Data from *Chemical Abstracts*, 1910–1960, are here presented showing the percentages of papers having a single author and those with two, three, and four or more. It seems evident that there has been a steadily accelerating change since the beginning of the century.

has accelerated steadily and powerfully, and it is now so large that if it continues at the present rate, by 1980 the single-author paper will be extinct.[22] It is even more impressive that three-author papers are accelerating more rapidly than two-author, four-author more rapidly than three-author papers, and so on. At present only about one paper in four has a multiplicity of three or more authors, but, if the trend holds, more than half of all papers will be in this category by 1980 and we shall move steadily toward an infinity of authors per paper. It is one of the most violent transitions that can be measured in recent trends of scientific manpower and literature.

One way of understanding this movement toward mass collaboration is to see it as a natural extension of the growth created by the constant shift of the Pareto distribution of scientific productivities. There is a continuous movement toward an increase in the productivity of the most prolific authors and an increase in the numbers of those minimally prolific. As we approach a limit in both directions, it is clear that something has to give. The most prolific people increase their productivities by being the group leaders of teams that can accomplish more than they could singly. The minimal group are in short supply, and we can hardly afford to let them grow until they reach that ripeness of producing significant papers on their own. By the creation of a class of fractional authors—that is, scientists who produce one nth part of a scientific paper—a much larger number of the minimal group is kept at the lower end of the distribution. One expects that as these individuals grow they will evolve into unit authors or better, but in the meantime the body of research workers is increased to meet demand. It is to some extent accidental that wartime organization and the advent of the big machine have occasioned the introduction of fractionality, without which we should have severe manpower shortage.

A more optimistic viewpoint to take is that the emergence of this class of sorcerer's apprentices partly solves the problem of organizing the lower-level scientists so that they can be directly related to the research life of the elite. There is nothing but a logical extension of that old familiar principle, the great professor with his entourage of graduate students, the sort of thing for which Rutherford or Liebig are well known. The great difference here is that the apex of the triangle is not a single beloved individual but an invisible college; its locale is not a dusty attic of a teaching laboratory but a mobile commuting circle of rather expensive institutions. R. E. Weston et al. have suggested that one might name such teams as the Dubna Reds and the Harvard M.I.T. Yankees, and give each player a rating.[23]

Because of this, one of the great consequences of the transition from Little Science to Big Science has been that after three centuries the role of the scientific paper has drastically changed. In many ways the modern ease of transportation and the affluence of the elite scientist have replaced what used to be effected by the publication of papers. We tend now to communicate person to person instead of paper to paper. In the most active areas we diffuse knowledge through collaboration. Through select groups we seek prestige and the recognition of ourselves by our peers as approved and worthy collaborating colleagues. We publish for the small group, forcing the pace as fast as it will go in a process that will force it harder yet. Only secondarily, with the intertia born of tradition, do we publish for the world at large.

All this makes for considerable change in the motivation of the scientist; it alters his emotional attitude toward his work and his fellow scientists. It has made the scientific paper, in many ways, an art that is dead or dying. More than this, the invisible colleges have a built-in automatic feedback mechanism that works to increase their strength and power within science and in relation to social and political forces. Worse, the feedback is such that we stand in danger of losing strength and efficiency in fields and countries where the commuting circuit has not yet developed. In short, now that we have achieved a reasonably complete theory of scientific manpower and literature, we must look to the social and political future.

4

Political Strategy for Big Scientists

In our analysis of the growth of science we have reached a basic understanding of normal exponential increase and distribution of talent and productivity. Now let us turn our attention to the abnormal—that is, to those things that do not follow the pattern. Without doubt, the most abnormal thing in this age of Big Science is money. The finances of science seem highly irregular, and since they dominate most of the social and political implications, our analysis must start here.

If the costliness of science were distributed in the same way as its productivity or excellence, there would be no problem. If the per capita cost of supporting scientists were constant, we should only spend in proportion to their number, so that the money they cost would double every 10 to 15 years. But in fact our expenditure, measured in constant dollars, doubles every 5½ years, so that the cost *per scientist* seems to have been doubling every 10 years. To put it another way, the cost of science has been increasing as the square of the number of scientists.

Since we know that in general the number of average scientists increases as the square of the number of eminent highly productive ones, we derive the frightening costly principle that research expenditure increases as the fourth power of the number of good scientists. It has already been estimated that the United States may possess enough talent to multiply the population of distinguished scientists by a factor of 5. Let us be conservative and envisage a future in which it is only tripled; we could reach this point quite some time before the year 2000. By then, according to the principle just derived, our expenditure would have multiplied by a factor of 81, and would thus be more than double our entire gross national product.

It seems incontrovertible that such an increase in the cost of science has been taking place. National research and development expenditures were about three billion dollars in 1950 and thirteen billion dollars in 1960—more than a doubling every five years. The 15 percent annual increase must be matched against a rise in the gross national product of only 3½ percent a year. At the present rate, science will be 10 percent of the gross national product as early as 1973. It is already in the region 2 to 3 percent, depending on definition.

Let us be optimistic and suppose that growth of the gross national product will continue, with no manpower shortage to impede the increase in the number of qualified scientists, and return to the question of whether the cost per scientist must also increase. Data from the federal agencies that now support so much research indicate clearly that the cost per project has been rising rapidly. The National Institutes of Health figures for average expenditure per project are $9,649 in 1950 and $18,584 in 1960, almost a doubling.[1] Johnson and Milton investigated the records of a wide range of research carried on in industry, universities, and government institutions and found that in a decade, although the total costs increased by a factor of 4½, the output of research and development no more than doubled.[2]

Basically it appears that as more and more research is done our habitual and expected increase therein is still needed but becomes more difficult to obtain. The result is that we offer more and more inducement by raising salaries, providing more assistance, and giving the researchers better tools for the job. This is essentially the Fechner law situation already described, the effect being proportional to the logarithm of the stimulus. However, apparently it is necessary to use up units of financial solidness about twice as fast as units of scientific solidness.

We may now inquire why the cost of research on a per capita basis and in terms of the gross national product seems to have remained constant throughout history until about World War II and only since that time has met with the new circumstance of an increase that keeps pace with the growth of scientific manpower.

Let me offer as an interpretation, not an answer, the suggestion that this is the cybernetic feedback that is now trying to decelerate science and bring it to a maximum size. This, I maintain, is the prime cause of the present logistic rather than exponential curve. This is the difference between Big Science and Little Science. But we cannot discover the reason until we have looked deeper into the world rather than the national situation, and into the motivation of the scientist.

Let us first examine the world situation, considering all the separate countries and the various bodies of science contained in them. For a first approximation these are normally distributed like the sizes of cities within a country, ranking from the few big ones down to the many small ones. There is uniform exponential growth, just as in cities. Just like the rank list of sizes of cities, as we watch it evolve through history, the order changes slightly, though the distribution remains stable. Over the years there is a change in which some countries alternately lead and lag behind others. It is a slow process, though the realization, as in the instance of Sputnik, can be a shock to the uninformed.

During the present century, world science has altered its national divisions almost systematically. Consider the figures showing the contributions of various countries to the production of scientific papers analyzed in *Chemical Abstracts* (fig. 4.1). At one end, the old and stable scientific culture of the British Commonwealth has been sensibly constant, and that of France has suffered

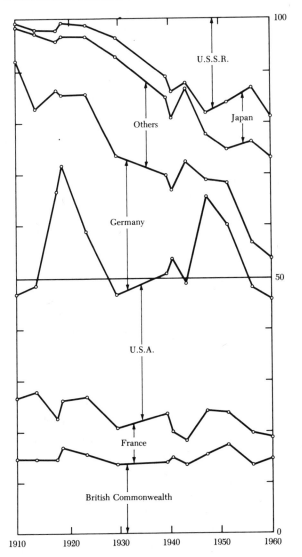

Figure 4.1. Percentages of World Total of Output of Papers in
Chemical Abstracts, by Countries, 1910–1960

a slight but steady decline. At the other end, the Soviet Union, Japan, and, indeed, all the minor scientific countries have spectacularly improved their world position from about 10 percent at the beginning of the century to nearly 50 percent now. In the middle, being squeezed by this expansion, are the two great chemical nations, Germany and the United States. Their combined share has declined from 60 to 35 percent, with the United States apparently absorbing a large part of the German share during both world wars, and Germany having shrunk to one-fifth of its original size.

Altogether, apart from the wartime winnings from Germany, the United States has approximately maintained its relative position. It has perhaps even made up the losses of France. Remember that this does not include the steady exponential increase at the world rate of a doubling every 10 years. The spectacular thing is not that the United States or any other country can maintain this rate and keep its position constant but that undeniably the Soviet Union, Japan, and the minor scientific countries have during the present century been able to exceed this world rate so that they have grown from nonentity to a near majority. Together they seem to have erupted into the scientific scene at a rate exceeding their normal quota of the scientific explosion by about 6 percent a year. Consequently, we now face a spectacualr decline in the traditional ability of big nations to form an absolute majority of science on their own. They are facing a Pareto-like distribution of smaller countries whose total bulk will soon outnumber that of the United States and Soviet Union.

Japan, the Soviet Union, and the United States all have present expenditures in the region of 2 to 3 percent of their gross national products. How is it possible that their relative productions can be shifting slowly but steadily? The likeliest explanation seems to be the steady increase in the cost of science, as society becomes saturated with this activity. A complementary effect is that it seems cheaper and easier than usual to make science explode into the "vacuum" of an underdeveloped country.

The present great activity in bringing science and its technologies to the little nation makes it worthwhile for us to look more closely at the conception and birth of a modern scientific and technological civilization. We must carefully distinguish the type of scientific explosion with which we are concerned, the emergence of a country relative to all others, from ther normal explosive change in which all countries are involved in proportion to their rank.

Most countries merely retain the same place in the hierarchy, complaining bitterly, like Alice, of being forced to run so hard to stay in the same place. The recent pronouncements in England of the Zuckerman Committee on scientific manpower seem to be like this. When a country decides that it can afford to let science grow only at the rate of the national economic expansion, and that the supply and demand of scientific manpower be allowed to tend to equality, this is tantamount to a suicidal withdrawal from the scientific race. Alas, it is the race that Britain, bereft of great resources of minerals or agriculture, should strive in above all else.

Since we seem to have a crystallization of science that tends to make the rich richer and the poor poorer, how does it happen that paupers occasionally turn into scientific millionaires? In one particular instance, history has provided the complete sequence of steps by which a nation suddenly emergent was able to explode more vigorously than the rest of the scientific world. An analysis of the data for Japan may stand as a prototype (fig. 4.2).[3] In 1869, at the beginning of the Meiji era, ca. 1868, Japan broke with tradition and invited the introduction of Dutch science, as our Western product was then called.

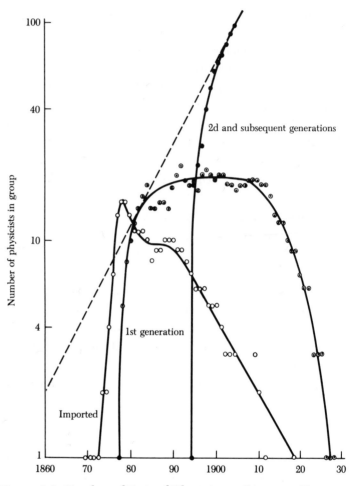

Figure 4.2. Number of Trained Physicists in Japan as a Function of Date

The "imported" curve counts Europeans and those trained in Europe. The next curve gives the numbers of their students. The third curve gives the number of Japanese students trained at home by Japanese teachers—this number grows as if it started from the original shock wave and grew exponentially to the present day, but only after a waiting period of about 15 years while the first generation was prepared.

Let us now trace the progress of but one science, physics. The first step was the importation of foreign science teachers from the United States and Great Britain, and the export of young Japanese students to foreign universities for advanced training. The shock wave of Western science hit the country abruptly and caused Japan's population of physicists to rise from 1 to 15 in only 6 years. By 1880, the shock wave had begun to die away, at first rapidly as the foreigners went home, then more slowly as foreign-trained students and teachers retired and died, so that this wave finished by 1918. But in 1880, when the imported curve was at its maximum, a new wave was rising rapidly; this was the first generation of Japanese students trained by the aforementioned foreigners and their disciples.

The first generation of students was a small group; there were 10 in 1880, and their numbers never rose above 22, reaching a stable balance between training and mortality. Later, around World War I, they began to decline noticeably in number, the last dying in 1928.

The second generation of students, those who were now being trained by Japanese in Japan, began in 1894 and rose to 60 graduates by 1900. Shortly thereafter growth settled down to the familiar exponential pattern, doubling every 10 years. This growth, continued without serious break or disturbance, led to the state of physics in Japan through the last war.

The total effect of the shock wave and the pulse of first-generation students is the inception of exponential growth. The eventual curve of growth, projected back, acts almost exactly as if it had sprung from the crest of that first shock wave, that is, as if it had started from 12 physicists in 1881. Note, though, that growth did not start immediately; there was a lag of nearly 20 years while the second generation prepared. It seems important that the steady state arose only with this crop of entirely home-grown physicists. The picture, however, is not complete. I have omitted the important fact that primary and secondary education had also to meet similar crises at similar dates. I have neglected the crucial point that the greatest difficulty of all was to decide the language of instruction. Not until the second generation could it possibly be in their native tongue, and then not before new vocabularies and new dictionaries had been compiled.

Simplifications notwithstanding, we now have the basic time scale and the shapes of the differential equations of scientific manpower in an underdeveloped country, and we have little reason to doubt that the case is typical. Most important is the lag while waiting for a generation, then the spurt that is faster than subsequent exponential growth. It is another instance of ripe apples falling easily from the tree. The apples here are not discoveries but potentially bright physicists in a country that has no physics. This aspect is the opposite extreme from a highly developed scientific country.

The explosion of science into an underdeveloped country can, then, if serious effort is made, be much faster than into one in which science is already established. In the case of the larger population masses of the world, the process is partly familiar, and partly a cause for grave concern.

The explosion into a vacuum is basically the reason why the United States, starting its scientific revolution much later than Europe, was able to proceed more rapidly to parity and then to outpacing. In exactly the same way, the Soviet Union, starting much later than the United States, has been able to expand at a greater exponential rate—perhaps a doubling in 7 years rather than 10. Similarly, now that China is emerging scientifically, as one can tell by the fact that we now routinely translate their chief journals as we have translated those in Russian for many years, one may expect it to reach parity perhaps within the next decade or two. The Chinese scientific population is doubling about every 3 years.[4]

Thus, for the great blocks of the world population we have a sort of automatic handicap race. The later a country starts its determined effort to make modern science, the faster it can grow. One may therefore suppose that at some time during the next few decades we shall see a rather close finish to a race that has been running for several centuries. The older scientific countries will necessarily come to their mature state of saturation, and the newly scientific population masses of China, India, Africa, and others will arrive almost simultaneously at the finishing line.

I maintain that this process is historically inevitable and that we must therefore preserve a sense of balance, and not panic during the forthcoming waves of Sputnik-like scientific advances by countries previously regarded as second-rate in high science and technology.

Let us now consider the distribution curve for the incidence of scientific talent in a country. Although we have no objective measure for the talent latent in an underdeveloped country, it is reasonable to assume it would be spaced out in the same way as in developed countries, with relatively few high talents and a number of lesser talents increasing more and more as the minimum qualification is approached. As we have seen, the general effect of increasing the total scientific population is to multiply the lesser talents faster than the highest ones which dominate the scene and produce half of all the scientific advance.

As long as the country is relatively undeveloped, the number of scientists will be too small to need much crystallization into groups and elites, for the entire body will consist of the cream that has risen to the top. As the country develops, crystallization into groups, into scientific cities, begins, as well as the diminishing relative return of first-class scientists. Effectively, more and more of the extreme tip of the tail of the distribution curve is used up, so there is a tendency to use up a longer and longer segment of the tail, to make scientists of those nearer and nearer to the average ability.

It must be emphasized that I am not saying that there is any lowering of the minimum standards for being a scientist. Merely, it seems that the effort to gain more scientists increases the number at the lower levels at a greater rate than it does those on the higher levels. Thus, although the number of highest caliber men can always be increased, this is done at the cost of the average standard. From the nature of this process it follows that at some stage between underdevelopment and high development one ceases to skim off the cream; society begins to have to work against the natural distribution of talents. Apparently it is inevitable that increased inducements and opportunities result in a smaller and less rich crop, albeit in enough of an increase in top people to make the process definitely worthwhile for a long time.

It is my thesis that the logistic decline from centuries of exponential growth takes place because we are scraping the bottom of the barrel in this way. At a certain point it may no longer be worthwhile to sacrifice so much to increase inducements and opportunities when the only result is a declining overall standard. It is a difficult thesis to maintain, for it might well be said that so long as first-class people can be produced, and so long as those already in being can be enabled to continue first-class work, it is always worthwhile to spend the effort and money. This, I think, ignores the general mechanics of any approach to logistic stagnation. The forces of growth, deprived of their customary booty, begin to apply themselves elsewhere, and a host of new troubles result. In scientific manpower, if we begin to scrape the bottom of the barrel, this shows up in several different ways. Perhaps the most apparent is an upsetting of the traditional and natural balance among fields and among countries. The tradition depended on the natural hierarchy of growths in the various areas; the new position, instead, largely depends on the hierarchy of forces associated with logistic decelerations.

In fields of scientific activity where once there was a natural sorting of people into various subjects according to their predilections and the caprice of opportunity and inspiration, society now offers various inducements and facilities designed to attract men to specific areas. Thus, the law of supply and demand begins to obey these different forces and the distribution changes just as effectively as if there were only a constant supply and a rapidly increasing demand. There is, indeed, the equivalent of a restricted supply of the highest talent manpower, so that there is increased competition to secure a high concentration of such talent in the midst of the decreasing density of it.

Thus, in this competitive situation, fields of high inducement gain on the low in a manner deviating from the tradition. In the United States and in England at present, it is easy to see competition between glamor subjects that get the men, and unglamorous ones that do not. There is an apparent falling away from the expected growth rates of graduate training in medicine, engineering, and education, which may be attributed to their maintenance in physics, mathematics, and astronomy.

Similarly, on the international scene, there appears to be a tendency for scientists to leave the countries where only minimal inducements and opportunities are needed to produce manpower and move to lands where in order to get the job done it must be made enormously attractive, notably the United States. The very internationality of science perhaps makes such movement more possible for scientists than it is for other classes of men. Thus, the countries of the British Commonwealth and Europe complain bitterly of their loss of high-talent manpower through emigration;

and we suffer the troubles consequent upon a flow from regions of scarcity to regions of plenty, and upon crystallization of the world's supply of the mother liquor of scientific manpower which causes such manpower to aggregate in already overflowing centers.[5]

Exactly the same process takes place among disciplines as among countries. Let us analyze it further in terms of the structure we have already found, the formation of small invisible colleges of a hundred or so men outstanding in each major field. As such a group develops into an integrated body, increasing its efficiency and ability to coordinate the activities of a large number of men and their projects, so the power of the group seems to increase even more rapidly than its size. Certainly, as we have seen, its expenses will grow as the square of the size. Thus, we have a phenomenon of positive feedback; the more powerful such a group becomes, the more power it can acquire. Unto him that hath seems to be given, and this automatically entails the deprivation of him that hath not.

At heart, the motivation that causes Turkish, Yugoslav, Canadian, and Brazilian scientists to emigrate to the United States is the same as that which induces potential students of medicine to try for a Ph.D. in physics. Big Science countries and Big Science subjects must offer additional inducement in order to maintain normal growth, and in so doing they tend to react upon Little Science and little countries.

This is as far as the present mathematical analysis of the state of science can take us, but it hardly begins to pose the most significant problems of the age of Big Science. We must next inquire within the disciplines of sociology and psychology for the explanation of the peculiar force of inducement and opportunity within the big processes of science. Having already noted that the motivation of the scientists and the role of scientific publication appear to be changed by the emergence of invisible colleges, we must examine this more closely.

If there are to be more scientists than just those who fall from the trees like ripe apples, willing to pursue their dedicated aim in any circumstances, inducements are necessary. During the past few decades in the Unites States and Soviet Union, and less so in the rest of the world, there has been a marked increase in the social status of the scientist.[6] Since he was needed, since there arose some competition, there was an automatic raising of general salaries and of the research funds and facilities commanded by the prestige and the cargo cult[7] of modern science. I do not at present argue about whether the returns justify the economic, social, or political outlay. Suffice it to note that each increase in prestige produces an undoubted payoff in increased results, but also a heightened competition that raises the stakes for the next round.

Once we are committed to paying scientists according to their value or the demand for their services, instead of giving them, as we give other dedicated groups, merely an opportunity to survive, there seems no way back. It seems to me evident that the scientists who receive the just and proper award of such recognition are not the same sort of scientists as those who lived under the old regime, in which society almost dared them to exist.

The matter would not be so worrisome if the only way to be a scientist was to be endowed with the appropriate talents; that is to say, if people became radio astronomers not by capricious circumstances or by drifting into the field, but because that is what they could do best. From modern studies of creative ability in the scientific fields it appears that general and specific types of intelligence have surprisingly little to do with the incidence of high achievement. At best, a certain rather high minimum is needed, but once over that hump the chance of becoming a scientist of high achievement seems almost random. One noted quality is a certain gift that we shall term *mavericity*, the property of making unusual associations in ideas, of doing the unexpected. The scientist tends to be the man who, in doing the word-association test, responds to "black" not with "white" but with "caviar." Such a schizoid characteristic is plain throughout the peculiarly esoteric scientific humor of Lewis Carroll, and in a thousand broadsheets and notices of laboratory bulletin boards.

I note, incidentally, that the reaction to this mavericity is what produces the also characteristic objectivity and conservatism of the good scientist, the resistance he exhibits toward discovery and mad associations found by himself and by others, the feeling that the other man must be wrong.[8] He is caught in a violent interaction of passionately free creation, on the one hand, and innate objective caution, on the other. According to MacKinnon, the highly creative scientist might almost be defined as the rare individual who can survive the acute tension between the theoretical and the aesthetic, the tightrope walker between truth and beauty.[9] Perhaps it requires an oddly stable schizophrenic trait, one made stable by becoming a scientist.

Big Science tends to restrain some expressions of mavericity. The emergence of collaborative work and invisible colleges, the very provision of excellent facilities, all work toward specific goals in research. They seem to exercise pressure to keep scientific advance directed toward those ends for which the group or project has been created. This is an old argument against the planning of research, and it always generates the response that we must be careful to give each man his head, to allow him to follow the trail wherever it might lead. But there is no way to ensure that the man will be motivated to follow the trail when prestige and status depend on recognition by the group.

When the prestige and status of an individual are sufficient, or when for some other reason the whole group can be induced to follow, it makes a breakthrough, a now familiar type of phenomenon that carries high additional status with it. Although there is therefore some group encouragement of the display of mavericity, it might well be that this applies only in special cases, and that we may now be wasting mavericity in other directions. Perhaps there is need for an active effort to provide a sufficiency of support for reasearch without objective, funding without project, means for study and status without obligation to subscribe to a specified goal, the sort of thing that is at present partially provided in institutes for advanced study and through high-status research professorships.

Returning now to the question of whether reward by Big Science produces a breed of scientist different from that of Little Science, let us look at the characteristics noted by all those who have sought regularities among groups of eminent scientists. Galton, one of the first investigators, noted that more than half of his group of distinguished scientists were the eldest or the only child in their families, and this proportion, much higher than average, has since been confirmed in several investigations. Galton noted also that an unusually high proportion of his subjects were very attached to one parent, most often the mother. In extension of this it has since been remarked that many of the great men of science lost one parent early in youth (before the age of ten) and became strongly attached to the other.[10] Case histories show that scientists often are lonely children who find it easier to relate to things than to people. In short, many peculiar characteristics of personality seem to apply to those who become scientists.[11]

I suggest that all these characteristics apply to people who became eminent in the days of Little Science, and that we do not yet have much inkling of whatever new characteristics have been elicited by the changes to the new conditions of Big Science.

Many of the personality traits found formerly seem to be consistent with the hypothesis that many scientists turned to their profession for an emotional gratification that was otherwise lacking. If this is true, be it only a partial explanation, one can still see how cataclysmic must be the effect of changing the emotional rewards of the scientific life. If scientists were, on the whole, relatively normal people, just perhaps more intelligent or even more intelligent in some special directions, it would not be so difficult. But since it appears that scientists are especially sensitive to their modes of gratification and to the very personality traits that have made them become scientists, one must look very carefully at anything that tampers with and changes these systems of reward. Any such change will make Big Scientists people of different temperament and personality from those we have become accustomed to as traditional among Little Scientists.

The new phase of science seems to have changed the system of gratification in two different ways. In one direction, we have introduced the reward of general social status and financial return where there was precious little before. In the other direction, we have caused the scientist to seek the approbation of his peers in a different way. The man of Maxwell's equations was something not quite the same as he of the Salk vaccine. Though according to the mythology a scientist is supposed to be eternally moved only by innate curiosity about how things work and what they can do, there is nowadays a slightly different social mechanism whereby a man is led to feel his personal inspiration and mavericity acknowledged among other men as having triumphed over ambient conservatism and caution as well as over the secrecy of nature.

If this is true at the highest level, it is also plain that in less stratified regions the invisible colleges and all the lesser groups confer status and the means of leading a good life. They exercise power, and the more power one has in such a group, the more one can select the best students, tap the biggest funds, cause the mightiest projects to come into existence. Such power does not, of course, represent any selfish lust on the part of the scientist. Society is supporting this structure and paying for it more and more because the results of his work are vital for the strength, security, and public welfare of all. With everything said to be depending on him, from freedom from military attack to freedom from disease, the scientist now holds the purse strings of the entire state.

I hope it is not overdramatic to compare the present position of our scientific leadership to that which has existed in other countries, and in this country at other times, among the groups that used to control the means of destiny. On occasions, military power has been overriding, and then the generals have been in control, behind each palace revolt and cabinet meeting. Elsewhere it has been finance and the control of capital that were the mainspring of the state and the implement of decision. Or, in legislative government, one has seen the vital place taken by men of legal training. In a democracy we are accustomed to finding the leadership taken by men emerging from all these fields that have been crucial to the world's destiny.

Until recently, the scientist, insofar as he played any useful role in matters of state, was a passive instrument to be consulted like a dictionary, to turn out the right answer on demand. Several scientists and nonscientists will believe that it is desirable to maintain in the face of all difficulty the proposition that the scientist should be "on tap but not on top." Without arguing the ethics of the case, one can point out that the positive feedback governing the power of scientists works against any such proposition. The increased status of scientists and scientific work makes them increasingly vital to the state and places the state increasingly in the position of putting technical decisions into technical hands.

However, I am arguing not so much for the assumption of control by scientists over matters within their technical expertise but rather that their new tendency to rise to the political front as representatives of a group of people who hold the purse strings of our civilization is to be encouraged. In a saturation economy of science it is obvious that the proper deployment of resources becomes much more important than expensive attempts to increase them.

In Great Britain and the United States very few of the senators, congressmen, members of Parliament, and active politicians— less than 3 percent, in fact—have had any training in science or technology. Among deputies in the Supreme Council of the Soviet Union, the figure now exceeds 25 percent,[12] and, though their machinery of government is very different from ours, I take this as an indication of the way our own future may lie.

In the old days of Little Science there was tremendous reaction against political action by scientists. They were lone wolves; they valued their independence; on the whole they liked *things* but were not very good at *people*. Their payoff was the approbation of peers, and they were not supposed to crave any sort of admiration from the man in the street or any social status within society. Whether they like it or not, they now have such status and an increasing degree of affluence. They have come within the common experience. When I first saw the comic-strip character of Superman, who had once looked so much like an all-American football player, metamorphosing before my very eyes into an all-American nuclear physicist, I felt that the old game was up, and that the President-after-next might well be an ex-scientist.

This is the credit side of the register which balances some of the other, not so good, changes already noticed in the first generation of new-style Big Science. The scientist is accepted by society and must shoulder his responsibility to it in a new way. The rather selfish, free expansion by exponentially increasing private property of scientific discoveries must be moderated when one is in the logistic state. Racing to get there before the next man might well be, in the long run, an impossibly irresponsible action.

It must surely be averred as a matter of principle that the country that has arrived at a full logistic maturity, saturated with science, must try to behave with maturity and wisdom; must give some guidance to the younger countries that are growing up around and gradually outstripping it in scientific superiority.

One of the things I think is happening is the maturing of a certain responsible attitude among scientists analogous to that which, in almost prehistoric times, moved physicians toward the concept of the Hippocratic oath. Contrary to popular belief, this happened not because doctors were unusually dedicated or pub-lic-spirited people but because they were all too easily held per-sonally responsible by their customers for poison, malpractice, and so on. The scientist has had a much harder time in arriving at this, for his customer has usually been the state rather than an individual. His guilt has been in the eyes of the world rather than in those of an individual. Here I refer not only to such matters as nuclear testing and fallout but also to a general question of what service science is rendering for the common good and for the improvement of man's higher understanding. Invisible colleges and groups now have the power to cast out their "poisoners and abortionists," and withdraw from them the old protective cloak of disinterestedness that was proper in the days of Little Science.

It is most heartening to find that on the whole, sundry much-publicized examples notwithstanding, the world body of scientists has been remarkably unanimous in political evaluations during recent years, and consistent in public action in an age of Big Science. Robert Gilpin's recent analysis of this consistency makes a most hopeful document.[13]

Scientists have hardly yet begun to realize that they hold in their own hands a great deal of power that they have hardly used. The ranks of senior scientists and key administrators of science have now swelled to the point where I think it will not be long before some of the good ones begin to enter politics rather more forcibly. We need such men, on the national scene and on the international scene. We need them for the internal reconstruction of the entire social fabric of science and for the external problems of science in the service of man.

It is my hope that in these lectures, beyond my own prides and prejudices in interpreting the data, I have shown that a whole series of annoyances and difficulties in scientific manpower and its literature are of a single process in which at last we find a change in the state of science the like of which we have not seen for 300 years. The new state of scientific maturity that will burst upon us within the next few years can make or break our civilization, mature us or destroy us. In the meantime we must strive to be ready with some general understanding of the growth of science, and we must look for considerable assumption of power by responsible scientists, responsible within the framework of democratic control and knowing better how to set their house in order than any other men at any other time.

5

Networks of Scientific Papers

This essay is an attempt to describe in the broadest outline the nature of the total world network of scientific papers. We shall try to picture the network which is obtained by linking each published paper to the other papers directly associated with it. To do this, let us consider that special relationship which is given by the citation of one paper by another in its footnotes or bibliography. I should make it clear, however, that this broad picture tells us something about the papers themselves as well as something about the practice of citation. It seems likely that many of the conclusions we shall reach about the network of papers would still be essentially true even if citation became much more or much less frequent, and even if we considered links obtained by subject indexing rather than by citation. It happens, however, that we now have available machine-handled citation studies, of large and representative portions of literature, which are much more tractable for such analysis than any topical indexing known to me. It is from such studies, by Garfield, Kessler, Tukey, Osgood, and others, that I have taken the source data of this study.[1]

INCIDENCE OF REFERENCES

First, let me say something of the incidence of references in papers in serial publications. On the average, there are about 15 references per paper and, of these, about 12 are to other serial publications rather than to books, theses, reports, and unpublished work. The average, of course, gives us only part of the picture. The distribution (see fig. 5.1) is such that about 10 percent of the papers contain no references at all; this notwithstanding, 50 percent of the references come from the 85 percent of the papers that are of the "normal" research type and contain 25 or fewer references apiece. The distribution here is fairly flat; indeed

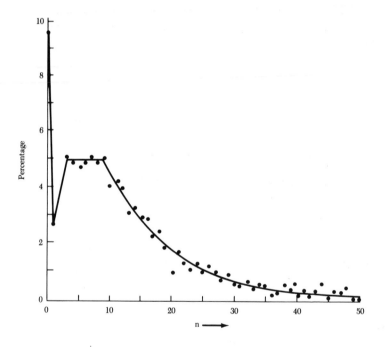

Figure 5.1. Percentages (relative to total number of papers published in 1961) of papers published in 1961 which contain various numbers (n) of bibliographic references. The data, which represent a large sample, are from Garfield's 1961 *Index* (2).

about 5 percent of the papers fall in each of the categories of 3, 4, 5, 6, 7, 8, 9, and 10 references each. At the other end of the scale, there are review-type papers with many references each. About 25 percent of all references come from the 5 percent (of all papers) that contain 45 or more references each and average 75 to a paper, while 12 percent of the references come from the "fattest" category—the 1 percent (of all papers) that have 84 or more references each and average about 170 to a paper. It is interesting to note that the number of papers with n references falls off in this "fattest" category as $1/n^2$, up to many hundreds per paper.

These references, of course, cover the entire previous body of literature. We can calculate roughly that, since the body of world literature has been growing exponentially for a few centuries, and probably will continue at its present rate of growth of about 7 percent per annum, there will be about 7 new papers each year for every 100 previously published papers in a given field. An average of about 15 references in each of these 7 new papers will therefore supply about 105 references back to the previous 100 papers, which will therefore be cited an average of a little more than once each during the year. Over the long run, and over the entire world literature, we should find that, on the average, *every scientific paper ever published is cited about once a year.*

INCIDENCE OF CITATIONS

Now, although the total number of citations must exactly balance the total number of references, the distributions are very different. It seems that, in any given year, about 35 percent of all the existing papers are not cited at all, and another 49 percent are cited only once ($n = 1$) (see fig 5.2). This leaves about 16 percent

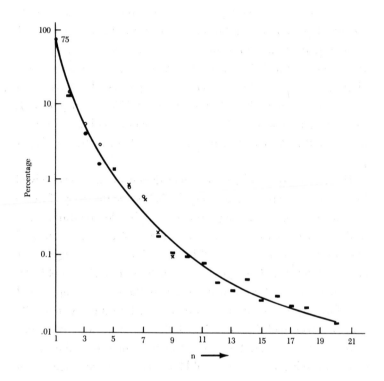

Figure 5.2. Percentages (relative to total number of cited papers) of papers cited various numbers (n) of times, for a single year (1961). The data are from Garfield's 1961 *Index*, and the points represent four different samples conflated to show the consistency of the data. Because of the rapid decline in frequency of citation with increase in n, the percentages are plotted on a logarithmic scale.

of the papers to be cited an average of about 3.2 times each. About 9 percent are cited twice; 3 percent, three times; 2 percent, four times; 1 percent, five times; and a remaining 1 percent, six times or more. For large n, the number of papers cited appears to decrease as $n^{2.5}$ or $n^{3.0}$. This is rather more rapid than the decrease found for numbers of references in papers, and indeed the number of papers receiving many citations is smaller than the number carrying large bibliographies. Thus, only 1 percent of the cited papers are cited as many as six or more times each in a year (the average for this top 1 percent is 12 citations), and the maximum likely number of citations to a paper in a year is smaller by about an order of magnitude than the maximum likely number of references in the citing papers. There is, however, some parallelism in the findings that some 5 percent of all papers appear to be review papers, with many (25 or more) references, and some 4 percent of all papers appear to be "classics," cited four or more times in a year.

What has been said of references is true from year to year; the findings for individual cited papers, however, appear to vary from year to year. A paper not cited in one year may well be cited in the next, and one cited often in one year may or may not be heavily cited subsequently. Heavy citation appears to occur in rather capricious bursts, but in spite of that I suspect a strong statistical regularity. I would conjecture that results to date could be explained by the hypotheses that every year about 10 percent of all papers "die," not to be cited again, and that for the "live" papers the chance of being cited at least once in any year is about 60 percent. This would mean that the major work of a paper would be finished after ten years. The process thus reaches a steady state, in which about 10 percent of all published papers have never been cited, about 10 percent have been cited once, about 9 percent twice, and so on, the percentages slowly decreasing, so that half of all papers will be cited eventually five times or more, and a quarter of all papers, ten times or more. More work is urgently needed on the problem of determining whether there is a probability that the more a paper is cited the more likely it is to be cited thereafter. It seems to me that further work in this area might well lead to the discovery that classic papers could be rapidly identified, and that perhaps even the "superclassics" would prove so distinctive that they could be picked automatically by means of citation-index-production procedures and published as a single *U.S.* (or *World*) *Journal of Really Important Papers*.

Unfortunately, we know little about any relationship between the number of times a paper is cited and the number of bibliographic references it contains. Since rough preliminary tests indicate that, for much-cited papers, there is a fairly standard pattern of distribution of numbers of bibliographic references, I conjecture that the correlation, if one exists, is very small. Certainly, there is no strong tendency for review papers to be cited unusually often. If my conjecture is valid, it is worth noting that, since 10 percent of all papers contain no bibliographic references and another, presumably almost independent, 10 percent of all papers are never cited, it follows that there is a lower bound of 1 percent of all papers on the number of papers that are totally disconnected in a pure citation network and could be found only by topical indexing or similar methods; this is a very small class, and probably a most unimportant one.

The balance of references and citations in a single year indicates one very important attribute of the network (see fig. 5.3). Although most papers produced in the year contain a near-average number of bibliographic references, half of these are references to about half of all the papers that have been published in previous years. The other half of the references tie these new papers to a quite small group of earlier ones, and generate a rather tight

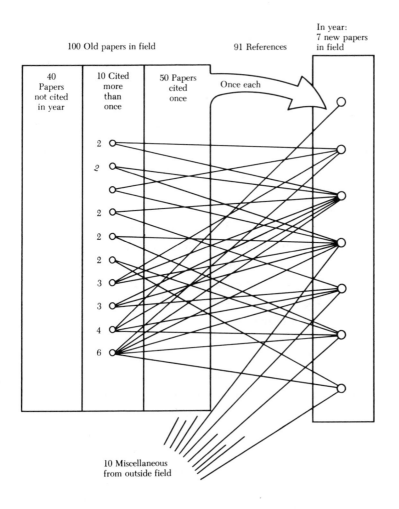

Figure 5.3. Idealized representation of the balance of papers and citations for a given "almost closed" field in a single year. It is assumed that the field consists of 100 papers whose numbers have been growing exponentially at the normal rate. If we assume that each of the 7 new papers contains about 13 references to journal papers and that about 11 percent of these 91 cited papers (or 10 papers) are outside the field, we find that 50 of the old papers are connected by one citation each to the new papers (these links are not shown) and that 40 of the old papers are not cited at all during the year. The 7 new papers, then, are linked to 10 of the old ones by the complex network shown here.

pattern of multiple relationships. Thus each group of new papers is "knitted" to a small, select part of the existing scientific literature but connected rather weakly and randomly to a much greater part. Since only a small part of the earlier literature is knitted together by the new year's crop of papers, we may look upon this small part as a sort of growing tip or epidermal layer, an active research front. I believe it is the existence of a research front, in this sense, that distinguishes the sciences from the rest of scholarship, and, because of it, I propose that one of the major tasks of statistical analysis is to determine the mechanism that enables science to cumulate so much faster than nonscience that it produces a literature crisis.

An analysis of the distribution of publication dates of all papers cited in a single year (fig. 5.4) sheds further light on the existence of such a research front. Taking from Garfield data for 1961 (see note 1), the most numerous count available, I find that papers published in 1961 cite earlier papers at a rate that falls off by a factor of 2 for every 13.5-year interval measured backward from 1961; this rate of decrease must be approximately equal to the exponential growth of numbers of papers published in that interval. Thus, the chance of being cited by a 1961 paper was almost the same for all papers published more than about 15 years before 1961, the rate of citation presumably being the previously computed average rate of one citation per paper per year. It should be noted that, as time goes on, there are more and more papers available to cite each one previously published. Therefore, the chance that any one paper will be cited by any other later paper decreases exponentially by about a factor of 2 every 13.5 years.

For papers less than 15 years old, the rate of citation is considerably greater than this standard value of one citation per paper per year. The rate increases steadily, from less than twice this value for papers 15 years old to 4 times for those 5 years old; it

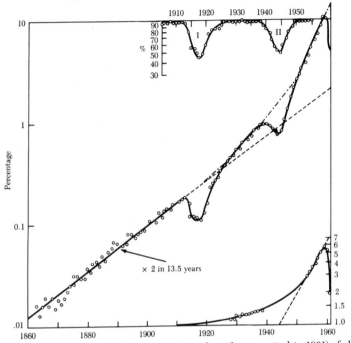

Figure 5.4. Percentages (relative to total number of papers cited in 1961) of all papers cited in 1961 and published in each of the years 1862 through 1961 (data are from Garfield's 1961 *Index*; see note). The curve for the data (solid line) shows dips during World Wars I and II. These dips are analyzed separately at the top of the figure and show remarkably similar reductions to about 50 percent of normal citation in the two cases. For papers published before World War I, the curve is a straight line on this logarithmic plot, corresponding to a doubling of numbers of citations for every 13.5-year interval. If we assume that this represents the rate of growth of the entire literature over the century covered, it follows that the more recent papers have been cited disproportionately often relative to their number. The deviation of the curve from a straight line is shown at the bottom of the figure and gives some measure of the "immediacy effect." If, for old papers, we assume a unit rate of citation, then we find that the recent papers are cited at first about six times as much, this factor of 6 declining to 3 in about 7 years, and to 2 after about 10 years. Since it is probable that some of the rise of the original curve above the straight line may be due to an increase in the pace of growth of the literature since World War I, it may be that the curve of the actual "immediacy effect" would be somewhat smaller and sharper than the curve shown here. It is probable, however, that the straight dashed line of the main plot gives approximately the slope of the initial falloff, which must therefore be halving in the number of citations for every 6 years one goes backward from the date of the citing paper.

reaches a maximum of about 6 times the standard value for papers 2½ years old, and of course declines again for papers so recent that they have not had time to be noticed.

Incidentally, this curve enables one to see and dissect out the effect of the wartime declines in production of papers. It provides an excellent indication, in agreement with manpower indexes and other literature indexes, that production of papers began to drop from expected levels at the beginning of World Wars I and II, declining to a trough of about half the normal production in 1918 and mid–1944, respectively, and then recovering in a manner strikingly symmetrical with the decline, attaining the normal rate again by 1926 and 1950, respectively. Because of this decline, we must not take dates in the intervals 1914–1925 and 1939–1950 for comparison with normal years in determining growth indexes.

THE "IMMEDIACY FACTOR"

The "immediacy factor"—the "bunching," or more frequent citation, of recent papers relative to earlier ones—is, of course, responsible for the well-known phenomenon of papers being considered obsolescent after a decade. A numerical measure of this factor can be derived and is particularly useful. Calculation shows that about 70 percent of all cited papers would account for the normal growth curve, which shows a doubling every 13.5 years, and that about 30 percent would account for the hump of the immediacy curve. Hence, we may say that the 70 percent represents a random distribution of citations of all the scientific papers that have ever been published, regardless of date, and that the 30 percent are highly selective references to recent literature; the distribution of citations of the recent papers is defined by the shape of the curve, half of the 30 percent being papers between 1 and 6 years old.

I am surprised at the extent of this immediacy phenomenon and want to indicate its significance. If all papers followed a standard pattern with respect to the proportions of early and recent papers they cite, then it would follow that 30 percent of all references in all papers would be to the recent research front. If, instead, the papers cited by, say, half of all papers were evenly distributed through the literature with respect to publication date, then it must follow that 60 percent of the papers cited by the other half would be recent papers. I suggest, as a rough guess, that the truth lies somewhere between—that we have here an indication that about half the bibliographic references in papers represent tight links with rather recent papers, the other half representing a uniform and less tight linkage to all that has been published before.

That this is so is demonstrated by the time distribution: much-cited papers are much more recent than less-cited ones. Thus, only 7 percent of the papers listed in Garfield's 1961 *Index* as having been cited four or more times in 1961 were published before 1953, as compared with 21 percent of all papers cited in 1961. This tendency for the most-cited papers to be also the most recent may also be seen in figure 5.5 (based on Garfield's data) where the number of citations per paper is shown as a function of the age of the cited paper.

It has come to my attention that R. E. Burton and R. W. Kebler have already conjectured, though on somewhat tenuous evidence, that the periodical literature may be composed of two distinct types of literature with very different half-lives, the classic and the ephemeral parts.[3] This conjecture is now confirmed by the present evidence. It is obviously desirable to explore further the other tentative finding of Burton and Kebler that the half-lives, and therefore the relative proportions of classic and ephemeral literature, vary considerably from field to field: mathematics,

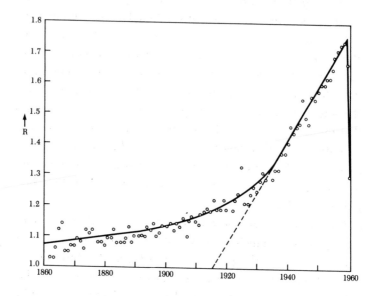

Figure 5.5. Ratios of numbers of 1961 citations to numbers of individually cited papers published in each of the years 1860 through 1960 (data are from Garfield's 1961 *Index*). This ratio gives a measure of the multiplicity of citation and shows that there is a sharp falloff in this multiplicity with time. One would expect the measure of multiplicity to be also a measure of the proportion of available papers actually cited. Thus, recent papers cited must constitute a much larger fraction of the total available population than old papers cited.

geology, and botany being strongly classic; chemical, mechanical, and metallurgical engineering and physics strongly ephemeral; and chemistry and physiology a much more even mixture.

HISTORICAL EXAMPLES

A striking confirmation of the proposed existence of this research front has been obtained from a series of historical examples, for which we have been able to set up a matrix (fig. 5.6). The dots represent references within a set of chronologically arranged papers which constitute the entire literature in a particular field (the field happens to be very tight and closed over the interval under discussion). In such a matrix there is high probability of citation in a strip near the diagonal and extending over the 30 or 40 papers immediately preceding each paper in turn. Over the rest of the triangular matrix there is much less chance of citation: this remaining part provides, therefore, a sort of background noise. Thus, in the special circumstance of being able to isolate a "tight" subject field, we find that half the references are to a research front of recent papers and that the other half are to papers scattered uniformly through the literature. It also appears that after every 30 or 40 papers there is need of a review paper to replace those earlier papers that have been lost from sight behind the research front. Curiously enough, it appears that classical papers, distinguished by full rows rather than columns, are all cited with about the same frequency, making a rather symmetrical pattern that may have some theoretical significance.

TWO BIBILOGRAPHIC NEEDS

From these two different types of connections it appears that the citation network shows the existence of two different literature practices and of two different needs on the part of the scientist. First, the research front builds on recent work, and the network becomes very tight. To cope with this, the scientist (particularly,

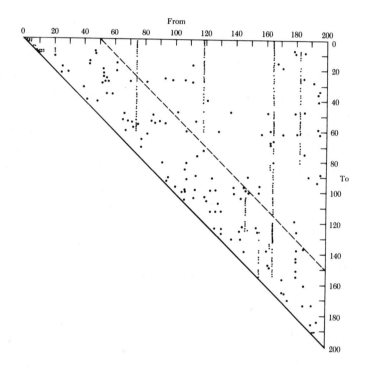

Figure 5.6. Matrix showing the bibliographical references to each other in 200 papers that constitute the entire field from beginning to end of a peculiarly isolated subject group. The subject investigated was the spurious phenomenon of N-rays, about 1904. The papers are arranged chronologically, and each column of dots represents the references given in the paper of the indicated number rank in the series, these references being necessarily to previous papers in the series. The strong vertical lines therefore correspond to review papers. The dashed line indicates the boundary of a "research front" extending backward in the series about 50 papers behind the citing paper. With the exception of this research front and the review papers, little background noise is indicated in the figure. The tight linkage indicated by the high density of dots for the first dozen papers is typical of the beginning of a new field.

I presume, in physics and molecular biology) needs an alerting service that will keep him posted, probably by citation indexing, on the work of his peers and colleagues. Second, the random scattering of figure 5.6 corresponds to a drawing upon the totality of previous work. In a sense, this is the portion of the network that treats each published item as if it were truly part of the eternal record of human knowledge. In subject fields that have been dominated by this second attitude, the traditional procedure has been to systematize the added knowledge from time to time in book form, topic by topic, or to make use of a system of classification optimistically considered more or less eternal, as in taxonomy and chemistry. If such classification holds over reasonably long periods, one may have an objective means of reducing the world total of knowledge to fairly small parcels in which the items are found to be in one-to-one correspondence with some natural order.

It seems clear that in any classification into research-front subjects and taxonomic subjects there will remain a large body of literature which is not completely the one or the other. The present discussion suggests that most papers, through citations, are knit together rather tightly. The total research front of science has never, however, been a single row of knitting. It is, instead, divided by dropped stitches into quite small segments and strips. From a study of the citations of journals by journals I come to the conclusion that most of these strips correspond to the work of, at most, a few hundred men at any one time. Such strips represent objectively defined subjects whose description may vary materially from year to year but which remain otherwise an intellectual whole. If one would work out the nature of such strips, it might lead to a method for delineating the topography of current scientific literature. With such a topography established, one could perhaps indicate the overlap and relative importance of journals and, indeed, of countries, authors, or individual papers by the place they occupied within the map, and by their degree of strategic centralness within a given strip. Journal citations provide the most readily available data for a test of such methods. From a preliminary and very rough analysis of these data I am tempted to conclude that a very large fraction of the alleged 35,000 journals now current must be reckoned as merely a distant background noise, and as very far from central or strategic in any of the knitted strips from which the cloth of science is woven.

6

Collaboration in an Invisible College

Many studies of the sociology of modern science and the communication patterns of scientists now agree that one of the dominant characteristics is that form of organization which has become known as the "invisible college." The name derives historically from a group of people in the mid–seventeenth century who later formally organized themselves into the Royal Society of London. Before that they had met informally, and distinct from the groups centered on Wadham College and Gresham College, the more visible colleges. They communicated by letter to gain an appreciative audience for their work, to secure priority, and to keep informed of work being done elsewhere by others. In its modern context of the organizational structure of Big Science, the term is not so specific, and unfortunately the definition and understanding of the term have varied considerably from writer to writer.

The basic phenomenon seems to be that in each of the more actively pursued and highly competitive specialties in the sciences there seems to exist an "in-group." The people in such a group claim to be reasonably in touch with everyone else who is contributing materially to research in this subject, not merely on a national scale, but usually including all other countries in which that specialty is strong. The body of people meet in select conferences (usually held in rather pleasant places), they commute between one center and another, they circulate preprints and reprints to each other, and they collaborate in research. Since they constitute a power group of everybody who is really somebody in a field, they might at the local and national level actually control the administration of research funds and laboratory space. They may also control personal prestige and the fate of new scientific ideas, and intentionally or unintentionally they may decide the general strategy of attack in an area.

These quite important phenomena are at present known only from personal histories and interviews, and so far as we know there has never been an objective analysis of an invisible college structure by any other means. It is relatively easy to interview a known "somebody" in a chosen research field, but opportunities are few to select a group of people that constitutes the greater part of a single invisible college. The basic difficulty of study is to capture and dissect out such a body. In what follows we present an outline and preliminary analysis of a likely specimen which was kind enough to agree to its capture and study, hoping that the information arising may be of value to those concerned with the analysis of scientific organization and communication and perhaps also to other groups of scientists. At this stage we have been at pains to preserve our primitive ignorance of the scientific content of the work of this group and also our lack of any personal knowledge of the participants in it; our hope is that some knowledge of the structure can be found first objectively and then checked later with the subjective data obtained from interviews.

The group investigated is Information Exchange Group No. 1 (IEG 1) on Oxidative Phosphorylation and Terminal Electron Transport, organized by the Division of Research Grants of the National Institutes of Health.[1] The group was established in February 1961 with 32 members. By June 1965 it had grown to 592 members, and another 6 information exchange groups in other specialties had been started by the same office.[2] Any scientist who is a bona fide researcher in the field in question may apply for membership in the group or may be nominated by other members. All members receive updated membership lists giving mailing addresses, and they all receive the numerous photocopied memos which are circulated. About 90 percent of these memos are preprints of papers which are eventually published, with or without change, and the remaining 10 percent are discussion of previous papers and an occasional technical or personal note intended for general circulation. Members may participate in this process which bills itself as a "continuing international congress by mail" by merely sending a fair typescript of their contribution to the central office. With no editorial intervention it is then photocopied and distributed so that scientist-to-scientist contact takes only a couple of weeks instead of the several months of delay attending formal publication. The total cost of the operation works out at about $125 of subsidy per member per year, and at about $.40 for each single copy of a memo. Though there has been much discussion of such techniques of shortcutting the normal channels of publication, particularly among the physicists,[3] we are not at this point concerned with the advantages and disadvantages of the IEG as a practical communication system, but only with the accidental by-product of a corpus of data which enables one to see something of the structure of the groups of people involved.

Our source data is a set of the membership lists and the memos which are circulated to all members. The membership list has been growing exponentially since its foundation, with a doubling time of about 13 months, so that in November 1965 there were 517 members. The number of memos has also been increasing exponentially, but with a doubling every 7 months, so that there was at the time of study about one memo a day on the average. The group of members is international—about 62% from the United States, 9 percent from the United Kingdom, 5 percent from Japan, 3 percent each from Australia and Sweden, 2 percent each from Canada, France, Germany, and the Soviet Union, and a remaining 10 percent from another 19 different countries. Compared with the average world state for all sciences, or even for the most active sciences, the growth rate is clearly an order of magnitude higher than one would expect, and the share of the United States probably rather disproportionately large. It would seem therefore that the group is still in process of spreading out to a fair coverage of all workers in the field. Some recent decline in the growth curves may be due to this process becoming effectively complete, though stable conditions have not yet been reached.

Our present report is based upon 533 available memos,[4] each of which bears on its byline the names of one or more authors. In all 555 different authors are named, and if we call each appearance of a name in a byline an *authorship*, there were 1,239 such authorships. It follows that each author had, in the five-year interval considered, about .96 papers and 2.23 authorships, there being overall about 2.32 authorships in the byline to each paper. The actual distribution of this multiple authorship is shown in table 6.1. The mode is clearly that of the two-author paper, and a side investigation was able to show that the distribution has not changed significantly over the five years.

Table 6.1
Distribution of Multiple Authorship

Authors per Paper	Number of Such Papers	Number of Such Authorships
1	114	114
2	230	460
3	123	369
4	45	180
5	14	70
6	4	24
7	2	14
8	1	8
Total	533	1,239

The trend in multiple authorship agrees closely with that cited by Beverly L. Clarke (2.30 authorships/paper),[5] but though similar in trend it is sensibly different from that found in this volume for chemistry and by Keenan for physics (1.80 authorships/paper).[6] It is quite clear that there are minor variations from field to field of science, and probably also variations in the national habits and conventions of who shall get his name on a paper, but on the whole the distribution in IEG 1 is in keeping with that expected. It has been shown by Hirsch and Singleton that the amount of multiple authorship in a field is closely related to the amount of financial support—government, foundation or private—given to the research producing these papers.[7] Presumably part of the origin of multiple authorship has a basis in the financial and economic as well as professional dependence of one author upon another. According to the data of Hirsch and Singleton the average number of authors per paper for nonsupported work in 1936–1964 was 1.16, while for supported work it rose from 1.38 to values of ca. 1.60 during the last three years of study when support had been most intense. The figure here of 2.32 would suggest that the field here studied must be highly subsidized. It is also worth noting that, as in other studies, the number of papers with n authors is proportional in the first approximation to $1/n-1$—except for the number of single-author papers (which should be twice as numerous). This law would hold for a random Poisson distribution of authorships with probability unity. If this, or anything like it were the cause of the distribution, it would point to the possibility that the group under study had systematically included too few single-author papers. Perhaps there is something in the nature of the subject matter or of the manner of organization that is prejudiced against a significant proportion of those who habitually publish without collaborators.

For a general analysis of the group of authors it is important to note that they are by no means coincident with the membership of the IEG 1. Though there were 555 different authors and IEG 1 had a membership of 517 scientists, only 231 names were common to both classes. There were therefore 324 authors who were not members of the group, and there were 286 members who had never helped to contribute a paper. Another side investigation showed that this breakdown, giving about 45 percent of the members as authors, was stable, it being the same in November 1964 as in November 1965, though presumably the founding fathers of the group were all considerable authors. The 324 authors who are not members of the group may reasonably be presumed to be the collaborators of authors who are members; in fact the ratio of all authors (555) to those who are also members (231) is 2.4 which is only slightly higher than the average number of authors per paper. The identity of the considerable number (286) of members who are not authors cannot be established with the means at present available; we shall hope that analysis of the references in the papers might indicate whether there are scientists in this field, and perhaps among this subgroup of noncontributing members, who are active in publication but opposed to this organization. Alternatively, members of this subgroup could be people who are interested but not active in the research front, or they could be people who are not now, nor perhaps have ever been, interested in this topic.

Focusing our attention now on the 555 authors who between them contributed the 533 memos, we next give the results of an investigation into their productivities and the extent to which they collaborated with each other. An index card was made up for each author named,[8] and on it were listed the papers in which he was listed as an author and the names of those other authors with whom he was listed for each paper. From each card the total number of papers and the number of different persons with whom he had been involved as a collaborator could be read. The results of this analysis are listed in figure 6.1, and also from this figure the marginal totals enable the distributions in productivity and amount of collaboration to be seen.

As discussed in the second essay of this book, the distribution of productivity alone follows the normal pattern so far as one can tell with the rather small numbers involved. A majority of the authors $(311/555 = 56\%)$ are known only from a single authorship of a paper. Since the aggregate of their authorships is only a

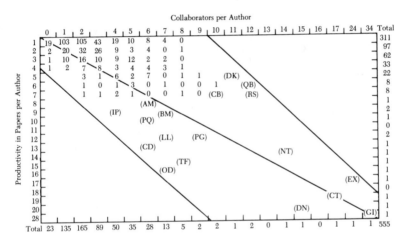

Figure 6.1. Numbers of authors in all categories of productivity and collaborativeness, and code names of the most productive persons. (Productivity and collaboration are well correlated; no author has many collaborators and low productivity, or vice versa.

quarter of the total number (311/1,239) it follows that in general they must be names whose single occurence has been on papers with several other authors. For larger productivities, the number of authors with their names on n papers is, in the first approximation, proportional to $1/n^2$ in the usual manner—a trend that has been constant for the last three centuries. At the other end of the productivity distribution there are as always a few very prolific authors. The top 30 authors who each contributed to 6 or more papers are responsible between them for another quarter (306/1,239) of the authorships.

The distribution of the number of collaborators per author has a well-marked maximum at two, corresponding to 3-author papers; a skewing of the distribution to the left makes this agree with the average of 2.32 authors per paper which has been already noted. At the top end of the distribution there are, this time, 17 people who collaborate a great deal, the minimum being with 8 people and the maximum being with 34 other authors.

Going now from the marginals to the main matrix of figure 6.1 it will be noted that there is a good correlation between the productivities and the amount of collaboration of the authors. The most prolific man is also by far the most collaborating. It is, of course, obvious that if multiple authorship is the rule a man will increase his number of collaborators with every paper he writes, but it is nevertheless surprising to find such a small scatter away from the main diagonal of the matrix. This implies that over the entire range of productivities there is little deviation from the main pattern of picking up about one new collaborator for each new paper written. There are in other words no people who display a marked tendency to stick to well-formed groups or to avoid previous colleagues. It is especially noteworthy that nobody who worked without collaborators or with only 1 coauthor succeeded in producing more than 4 papers in the five-year period, whereas everybody with more than 12 collaborators produced 14 or more papers in the same time.

The natural inference from this is that there exists a core of extremely active researchers and around them there is a large floating population of people who appear to collaborate with them in 1 or 2 multiple-authorship papers and then disappear not to be heard from again. We investigate this phenomenon in two ways—first by the calculation of *fractional productivities,* and second by the investigation of the way in which authors are grouped by the sets of all people who have ever collaborated with each other.

We define fractional productivity as the score of an author when he is assigned $1/n$ of a point for the occurence of his name among n authors on the byline of a single paper. Thus a man with one paper of which he is the sole author, a second of which he is one of two authors, and a third in which he is one of five, will have a fractional productivity of 1.7 and a full productivity of 3 papers. On the average it happens that the fractional productivities as so calculated are about half of the full productivities for most authors. More than two-thirds of the population of authors (380/555) had fractional productivities less than unity by this calculation, so confirming the existence of this large floating population of lightweights.

The use of fractional productivities provides a breakthrough in understanding the nature of the distribution law of productivity. For theoretical reasons, as shown in essay 2, it is to be expected that the logarithm of productivity should be normally distributed in a population. This is difficult to test with the integral values of full productivity, since the number of people with the minimum of unit productivity is already usually more than half the total population. Now, using fractional productivities it turns out that the expected law is quite well obeyed; a plot on logarithmic probability graph paper showing that those with less than a unit

of fractional productivity are on a normal curve continuous with
and a reasonable extension of that for people with fractional productivity greater than unity, the same people being in much the
same rank order at the high end of the distribution. Interestingly,
the median of the distribution is at a little less than a fractional
score of .5 or half a paper per man, and the standard deviation is
the same as for full productivity. We suggest on this basis that
part of the social function of collaboration is that it is a method for
squeezing papers out of the rather large population of people who
have less than a whole paper in them. Since the probability that
a man can double his score of papers is known to be about 1 in 4,
it follows that for every person with a whole paper in him there
may (if the curve can be legitimately extended) be 4 people with
only half a paper each. If collaboration is forbidden one gets only
1 paper, but if half papers are permitted by society, as they now
are, an extra 2 papers can be produced by suitable collaboration.
If quarter papers are permitted another 4 extra papers will be
forthcoming, and presumably the process might be continued
indefinitely. A marked bending of the distribution line in figure
6.2, however, indicates that for the smallest fractional productivities there are fewer authors than one would expect, so that there
seems some reluctance of society to accept its full quota of those
with the present minimum of one-eighth of a paper each—there
are only 4 such people in the sample and a normal distribution
would contain about 60.

In some fields of science there is a strong convention, different
from field to field, about the order of the names in the byline of
a paper. In theoretical physics it is invariably alphabetic, in some
biosciences it is in order of seniority of the author or in the
magnitude of his personal contribution to the collaboration. To
test for this effect a preliminary investigation was made of the
distribution of first-named authorships. Much more variation was

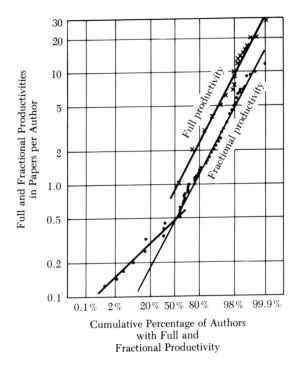

Figure 6.2. A cumulative percentage distribution of authors on full and fractional productivity. (The use of fractional productivities shows that the distribution of productivity is in good agreement with the theoretical expectation that the logarithm of the productivity is normally distributed in the population.)

found than with the comparison of fractional and full productivities. The top three men were highest equally in all varieties of productivity, but below this level there were a number of high producers on full and fractional counts who rated lower in first authorships than those who were quite low on the other two counts. The convention seems to vary considerably, therefore, from man to man, and further investigation is needed.

Having now confirmed the existence of a core and a floating

population, it remains to consider their relation to each other. To this end, the deck of index cards of authors was sorted manually so as to place together each author with those who had collaborated with him and also with those who had collaborated with his collaborators, etc. Thus, if 2 authors had collaborated on a paper, and then 1 of them had gone on to publish a second paper with 3 new additional collaborators, all 5 would be in a group that had produced 2 papers, one with 2 authors and another with 4. The result of the investigation is shown in table 6.2. At the top of this table is a line indicating that 23 of the 122 groups consisted of a single individual each, an author who collaborated with nobody else, and that these authors produced a total of 30 papers. At the bottom of the table is a line showing that the largest group of all contained 77 authors who collaborated in many different combinations in producing 117 papers.

The five largest groups account in all for about a third of the total population of authors, and each group contains one or more individuals whose record is high in productivity score and in number of collaborators. Typically, each group contains a small number of very active people and a large number of individuals who have collaborated only on a few papers. For example, in the largest group of all there are 6 of the 77 authors whose names appear on 28, 15, 14, 12, 10, and 9 papers, respectively, for a total of 88 of the 271 authorships within this group. In this same group 55 of the authors (71 percent) have their names on 3 papers or less and produce a total of 99 authorships. Similarly, in the second largest group, the top authors have 19, 17, and 12 papers each for a total of 48 of the 161 authorships, while 30 of the 58 authors have their name on a single paper only.

Perhaps the most striking feature of this part of the investigation is the finding that separate groups exist in what would otherwise appear to be a single invisible college. One would naturally expect

Table 6.2
Characteristics of Groups of Authors Related to Each Other by Collaboration

No. of Authors in Group N	Code Name of Big Man	No. of Such Groups G	Total Authors in Such Groups M = NG	No. of Papers from Group P	No. of Authorships A	Full Productivity P/M	Fractional Productivity A/M	Multiplicity of Authorship A/P
1		23	23	30	30	1.30	1.30	1.00
2		30	60	45	85	.75	1.42	1.89
3		27	81	56	128	.69	1.58	2.29
4		13	52	35	86	.67	1.65	2.46
5		7	35	32	80	.91	2.29	2.50
6		7	42	28	82	.67	1.95	2.93
7		7	49	42	105	.86	2.14	2.50
8		3	24	30	72	1.20	3.00	2.40
9	BM	1	9	7	19	.78	2.11	2.71
14	RS	1	14	12	36	.86	2.58	3.00
31	DN	1	31	32	84	1.03	2.74	2.63
58	CT	1	58	67	161	1.16	2.78	2.40
77	GI	1	77	117	271	1.52	3.52	2.32
	Totals	122	555	533	1,239			
					Average values	0.96	2.23	2.32

that authors who do not collaborate or those who have published a memo with only 1 or 2 other people would remain as isolated monads, pairs, and triads in such a study of groupings. One would, however, reasonably suppose that with authors who collaborate a great deal and with a constant flow of people each person would rapidly become connected, at some distance or other, with every other person. This does not happen, and there are at least five major noninterconnecting groups and probably several more of slightly smaller size.

A preliminary check of the listed locations of authors shows that each group is centered on, but not confined to, the institution or working area of its leading members. The biggest group is gathered around the largest research institute for this field, directed by their most active producer and collaborator who is moreover the chairman of this IEG 1 and a founder of the whole IEG apparatus as well as an editor of an important journal in the area. Another group exists in several Japanese institutes and universities.

Another noteworthy result of this phase of the study is that although the number of authorships per paper does not seem to vary significantly with the size of group (except that it must be unity for single-author papers and far below average even for 2-author papers) there is a striking variation of the productivities from group to group. In terms of the number of complete papers produced per man, the value is much higher than average for the largest groups and also for those who work alone. It is smaller than average for all groups of sizes in between. The effect is even more striking for fractional productivities of the groups measured in terms of the number of authorships per man; in this case the productivities are seen to increase rather regularly with the size of the group, the largest groups having values much higher than the smallest groups. Though the effect is large, its interpretation is most difficult. If the good of science be measured in terms of papers produced, the gain can be maximized by having scientists work either singly and without collaboration, or in very large interrelated groups. Medium-sized groups lower the amount produced. If, on the other hand, the gain to the author be measured in terms of the number of authorships he collects, then he will increase his stock by moving to the largest available group and decrease it drastically by working on his own. Clearly both goods are well served only by the largest groups, and it is therefore for this reason that the invisible college groups may exist. Exactly why this should be so is however a matter which is still dark to us.

Among the most interesting further problems which arise from this research is that of determining whether the palpable invisible college which has been studied is in fact several different, relatively unconnected, separate groups. Partly this could be determined by interviews, partly it could be checked automatically by seeing whether the references in papers of one group tended to cite that group exclusively or involved one or more of the other groups, too. Further, this analysis of references might well reveal the existence of any body of notable contributors to the research front of the literature who were neither known authors from this study nor nonauthor members of the IEG 1 group. A very cursory preliminary investigation indicates that the amount of self-citation by a group is heavy—about a quarter of all references rather than the one-tenth that is normal to scientific papers—but that references to the other groups also occur. It also indicates that about one-third of the references are to authors outside the IEG 1 authorship rolls. The data is, however, far too fragmentary for even tentative conclusion beyond the fact that linking authors by their involvement in collaboration provides a technique which is likely to agree with citation linkages and with the subjective estimates of what may constitute an invisible college.

The implications of this study are considerable for analyzing the social life of science and the nature of collaboration and communication at the research front. Not only have we indicated that the research front is dominated by a small core of active workers and a large and weak transient population of their collaborators, but we point the way, in conclusion, to the possibility that it is by working together in collaboration that the greater part of research front communication occurs. Perhaps the recent acceleration in the amount of multiple authorship in several regions of science is due partly to the building of a new communication mechanism deriving from the increased mobility of scientists, and partly to an effort to utilize larger and larger quantities of lower-level research manpower. If this is so, then the conventional explanation of collaboration, as the utilization of many different skills and pairs of hands to do a single job otherwise impossible to perform, is woefully inadequate and misleading.

7

Measuring the Size of Science

In the atomic and space era, science is rapidly becoming far too important to be left to the scientists. Part of the reason for this is that scientists have displayed incredible ingenuity over the ages in furthering their self-interest. From the time of Archimedes onward, they have been able to demonstrate conclusively to any government that maximum support of every need of scientific work was essential for the military and economic security of the state. Unfortunately, the demands of scientists now begin to exceed the possibilities of support, the pinch being felt first in the largest and most developed countries. We therefore begin to have a problem of "overdeveloped countries," where one must somehow learn to say no to at least some of the reasonable demands of the scientific community.

Another reason for the control of science passing out of the hands of the scientists is that there has begun to accumulate a considerable body of conventional wisdom and technical knowledge about the organization and the mechanisms of science in the structure of a nation's industrial, economic, and educational life. The time is passing when an experienced senior scientist or a managerial civil servant could pick up fairly quickly the small amount of previous literature and experience. Essential material of heavy scholarly content now appears at the rate of several books

Read before the Israel Academy of Sciences and Humanities on Februrary 11, 1969. The tables were contributed by Dr. Eugene Garfield, Institute for Scientific Information, Philadelphia.

a year, and a few years ago there was published a bibliography of bibliographies of "Science Policy Studies," which lists a few dozen monthly and annual bibliographies of the journal and thesis literature in this field.

I believe we are getting to the point where there must arise a fairly hard, respectable, and useful academic discipline that will do for science what economics does for the economic life of nations. We can no longer leave it to the "businessmen" of science, but need a Keynesian type of theory, partly for its use in policy decisions and in meeting crises before they burst upon us, and partly because we need to understand the machinery that makes science act the way it does and grow the way it grows. It is the business of sociologists to be knowledgeable about things that are important to society, and it is not necessarily the business, nor does it even lie within the competence, of natural scientists to turn the tools of their trade upon themselves or to act as their own guinea pigs.

Since the 1920s and 1930s, when this sort of "Science of Science" came into being,[1] it has been evident that the essential difficulty was in devising some reasonable measure of scientific effort or output. The monetary inputs to science and technology have always been clear and rather readily available, but they have defied all efforts to discern any regular pattern or law in the mass of data. Similarly, from the beginning it has been clear that almost all the figures available for scientific and technical manpower would not stand up to international comparison and, in short, no single measure of science could be related to any other measure of science.

This dismal state of affairs is due to two separate causes, as we can see now. For one thing, science is such a competitive activity that its manpower is distributed in a form resembling a very steep pyramid, with very few people of the highest caliber at the top and rapidly increasing numbers as competence decreases. In comparing two different countries, we may observe that a very small change in the level of qualification can cause very large changes in the size of the scientific and technical manpower. For another thing, we now begin to recognize that the scientific and the technical manpower groups do not form any sort of unity nor even a continuum. They are quite different social groups, comprising, on the one hand, the people who create new knowledge—the scientists, theoretical and applied—and, on the other, those who make new things, new chemicals, new machines: the engineers and technologists. The dividing line is by no means clear; there are many people with scientific and technical skill and training who make nothing new, adding neither to our knowledge nor to our artifacts, but work, with their know-how, well behind the research front.

In addition to these difficulties, one has the problems that attend the analysis of any sort of international demographic and economic data. Besides differences of definition, there are differences of purpose. Sometimes figures are designed to be as large as possible, sometimes as small as possible. Sometimes a country is at pains to show as much expenditure as it can in this direction, and, at other times, the figures are reduced to a minimum. In the United States and some other countries, research expenses in the most scientifically sophisticated industries carry a certain tax advantage, and it is important, therefore, that quite large expenses, particularly in the development of aircraft and electronic equipment, are included under the heading "Research and Development." Indeed, these categories dominate the whole national expenditure so that conventional figures for the investment of the nations in Research and Development primarily reflect these expenses, giving almost no indication as to whether one country is actually spending more or less than another on scientific and technical investigations, outside of what is essentially part of the cost of making aircraft and electronic equipment.

The difficulties I have mentioned are considerable, but, fortunately, there is one aspect of science that enables us to circumvent all these troubles. Science has an intrinsic quality of universality and internationality—one might even call it "supranational," in a remarkably strong sense of this word. Scientific knowledge is such that one cannot tell, apart from the name, whether the Planck Constant was discovered by a black man or a white, a Jew or an Arab, a German or a Mexican, a rich or poor man, a socialist or capitalist. Because of the utter impersonality of scientific creative knowledge, we have the paradox that Robert Merton has pointed out: one can only secure this private intellectual property of discovery and creativity by open publication.[2] The more open the publication, the more secure the private property. It is for this reason, I think, that there has come about a certain international constancy in what constitutes a publishable and good scientific paper in a good journal in any particular field. It might differ from field to field; I suspect, for example, that in every country it is rather easier to publish a paper in chemistry than one in physics. Clearly, too, papers differ enormously in their worth, the most important ones being a tiny fraction of the total. It seems, however, that the lower limit of what is publishable in the good international journals is part of a rather effective international general agreement—almost to the point of conspiracy—in order to keep the operation workable.

Because of the constancy in the level of papers, most people concerned with the measurement of scientific outputs have been reduced to using counts of paper, the men who write them, and the discoveries contained in them. The accepted customary technique is to point to the availability of great masses of such data collected for other purposes (such as scientific bibliography) and to the attractiveness of using these by-products of scientific activity as a noninterfering and silent way of measurement. One then points out that it is inevitably much more convenient to make a measurement, which is relatively easy, than to determine exactly what it is that has been measured. After this, one uses the numbers as one pleases.

A word of advice is necessary, especially to social scientists, about the way in which these numbers are to be used. I suggest that they are not to be taken merely as empirical data to be correlated and projected in the usual way. I think that, instead of the usual cookbook statistical methods, the paradigm should be the methodology of astrophysics in the days of Jeans and Eddington. One took whatever data came from the heavens, since experiment and measurement were out of the question, and then one tried to find suitable laws and models that would explain why these observations had the form and the order of magnitude that they did.

Thus, if it is stated that a country spends 4 percent of its gross national product (GNP) on Research and Development (R&D), or if one says that Israel has 1 percent of the world's population of nuclear physicists, or that one-fourth of all scientific discoveries are rediscoveries, we can only admit that such a statement is meaningful if we know what we should expect and why. One does not require the accuracy and technique used in testing Boyle's Law, nor the laboriousness of regressions and correlations; one needs the same rough and ready methods, initially, as in asking how many stars there should be in a galaxy.

The chief limitation in the counting of papers and authors and discoveries is that one needs published listings of quite large selections of papers having a reasonable degree of uniformity and comprehensiveness. For the earlier days, one has complete listings of all scientific journals through the eighteenth century and fairly good directories of the world's learned periodicals since that time. For the eighteenth century there is available, thanks to Reuss,[3] a complete, but not easily countable, bibliography of all the scientific and medical literature. For the nineteenth century there exists the excellent and very complete *Royal Society's Catalogue of Scientific Papers, 1800–1900,* but that again is countable only by sampling and guessing. For the more modern period, one has excellent sources for particular sciences in the comprehensive statistics available in *Chemical Abstracts* and *Physical Abstracts,* and a few similar publications that attempt comprehensive coverage of all significant publications in specific fields.

With the help of such sources, our knowledge of the statistical regularities of the measurements of scientific output has been considerably advanced. We know now that the average productivity of scientists—the number of scientists who write one paper, two papers, three, etc., in a given interval of time—does not vary from country to country very much, and hardly at all from century to century, since the invention in the seventeenth century of the scientific paper right up to the present day. This is a helpful confirmation of our supposition that the quantum of publication has stayed conveniently constant in the aggregate. It serves, also, to corroborate the usefulness of the definition that a scientist is any person who has ever published a scientific paper. In terms of this definition one can convert to almost any other that may be suggested, particularly one with similar but more stringent qualifications—for example, that a scientist is any person who has published an article in an international journal at least once within the last two years.

At all events, it has long been known that, almost irrespective of the exact definition, we are led to the interesting discovery that these measures of scientific production and manpower have all increased exponentially with a quite impressive and fatalistic regularity. They grow at the prodigious rate of a doubling every 7 to 10 years, depending on exactly what one is measuring. It is a rate of growth that is much faster than that of all the nonscientific and nontechnical features of our civilization. This leads, of course, to the now well-known conclusion that 90 percent of all the scientists that have ever existed are alive now. It also points to the fact that most scientists are young, and that, therefore, most scientific discoveries are being made by young men. It is perhaps worth emphasizing that all these statements are based on the exponential character of the growth alone, and are no more true today than they were a century or even two centuries ago. If science appears to be especially flourishing today, this is because of quite different factors. Indeed, many of these phenomena are not due to an increase in rate but to quite the opposite trend. In the overdeveloped countries, it is becoming difficult to keep up the pace to which science has become accustomed.

In addition to the impressively exact and regular laws governing the distribution of scientific productivity and the exponential general growth, there are several other well-marked regularities that serve to confirm that the counting of papers and authors is a useful procedure. A number of experiments have been made in assessing papers by their impact in terms of the number of citations they receive. There have also been experiments in apportioning the credit for papers among all the authors on the byline system instead of awarding an entire contribution to each, as is perhaps more customary. In both cases, one can show that the effect on the general law is negligible, though naturally the score of any particular author may be much affected.

In other directions, we have been able to chart significant differences in the network structure linking paper to paper in different scientific fields, and there has been some interesting success in the analysis of the process of multiple discovery, when two or more authors at the same or even at different times discover and publish essentially the same thing without any conscious knowledge of the other person's work. It can be shown that such duplication is normal, and that it is completely fortuitous, following a Poisson distribution. About a quarter of all discoveries are rediscoveries; science is a highly redundant process, and that may well be part of the essence of its self-checking character.

Perhaps the most interesting, and undoubtedly the most important, numerical data that we have relate to the way in which science is distributed among countries and fields of science. The basic question of science policy is, after all, how much money to spend on each of the various scientific and technical activities. Within the last few years there have been published breakdowns by country of publication for all the many thousands of papers a year that are noted in the key journals *Chemical Abstracts* and *Physics Abstracts;* in the latter case, there exists also a breakdown by field of physics as well as by country of publicaton.[4] The first notable general finding is that the shares of the countries are similar for chemistry and physics, even between the various segments of physics.[5] One might, perhaps, have supposed that there would be considerable variaton from country to country, but there are very few irregularities, and those that exist are almost immediately explainable by some well-known peculiarity. A second finding is that the share each country has of the world's scientific literature by this reckoning turns out to be very close—almost always within a factor of 2—to that country's share of the world's wealth (measured most conveniently in terms of GNP). The share is very different from the share of the world's population and is related significantly more closely to the share of wealth than to the nation's expenditure on higher education.

The reason for the approximate equivalence is basically simple. All other things being equal, one would suppose that the scientific size of a nation must be proportional to its population: two equal countries added together would produce a double-sized country in science as well as in population. It also seems reasonable to assume that the scientific size must be related to some function of the wealth of the country as might be measured by its per capita income. Being proportional to both these factors, the scientific size is determined then by population multiplied by some function of per capita wealth. Now this product must also be additive, so that the only simple function that can be used is the per capita wealth itself, and by multiplying it by the total population, one gets back directly the total wealth. This is most conventionally measured by GNP, the roughness of the present data making it useless to distinguish between the GNP at factor cost or any other reasonable variety expressed in constant currency. An alternative statement of the proposition, in terms that may be more acceptable to some, is that for the countries of the world the per capita activity in science correlates well with the per capita wealth.

To give a particular example, the United States publishes about one-third of the world's physics and chemistry as well as one-third of the astrophysics, and gets about one-third of the big prizes and discoveries—and has also about one-third part of the world's wealth. Its share is not anything near to 6 percent, which is its share of the world's population, nor to the much more than one-third which it has of the world's military expenditure or university population. To take another telling comparison, Canada and India each possesses about 2 percent of the world's GNP and a similar amount of its science, even though the population of India is twenty-five times that of Canada.

The biggest fault in all the previously available data on these lines has been that it was classified by country of publication rather than by the place in which the scientific work was actually performed. This appeared to have relatively little effect on the expected figures for most of the larger countries, each of which has several internationally used journals. There were, however, obvious distortions, such as that due to the publication in the Netherlands of several physics journals of international rather than national scope. It also had the severe disadvantage of neglecting almost entirely the contributions of those nations that were smaller than the "Big Ten," and of publishing the greater part of their good papers in journals based in the larger countries. It was, therefore, not clear from the old figures whether there was in fact a sort of scientific desert in which hardly any scientific work was carried out.

Thanks to a new by-product of the constant quest of the scientist for better bibliographic tools, we can now correct this view and provide for the first time a set of figures for all the smaller scientific countries and towns, including those of Israel, which happen to be more extreme than any other case known. The new tool we have used recently is the *International Directory of Research and Development Scientists, 1967*, which is published by the Institute for Scientific Information in Philadelphia, who publish also the *Science Citation Index*, which has also been tremendously valuable as a source of statistical data. The new *Directory* lists geographically by country, state, and town, as well as institution, the name of each scientist whose name was placed first among the authors of papers listed in *Current Contents* during the year 1967. The journal *Current Contents* covers, of course, a wide selection of all the world's major journals in all fields of science and technology, including medicine, and we know from tests that it covers somewhere between 80 and 90 percent of the impact value, as measured by citations, of the journal literature.

The biggest fault of the new data is that, by listing only first authors, one loses a lot of names, and this occurs to a particularly large extent in those countries and sciences where one has the institutional and hierarchical convention of an institute chief or professor whose name automatically comes first. As a result of the lack of an international convention, the figures for such countries may be reduced by as much as a factor of 2 below that for countries where the names are ranked in order sometimes of alphabet, sometimes of merit, sometimes of status, and sometimes by none of these principles. The data do, however, supply for the first time fairly reliable figures for towns and countries for which no previous count has been sufficiently universal and large. The numbers are, of course, valid only on a relative scale and have little absolute significance, because only a fraction of the producing authors will actually publish within any given year, and the location given is always that noted in the paper as the official address from which reprints might be obtained. It is, however, difficult to find a more adequate definition of the country of production of science that can be applied so automatically.

The result of a computer count of the first year's production of this index, including several faults of proofreading and debugging, which will improve no doubt from year to year, are now given in abbreviated form in the attached tables. From the list of countries (table 7.1) it can be seen that 90 percent of the world's science resides in the top 14 nations, and that 40 nations in all account for all but 1 percent of the world's contribution. Israel is nation number 17, with a scientific population comparable in size with Poland and Hungary, or with the cities of Leningrad, Los Angeles, or Cambridge, Massachusetts (see table 7.2). Alternatively, it has rather more scientists than the whole of Latin America and many more than the whole of Africa. Israel's largest scientific city, Jerusalem, is similar in size to Rome, Manchester, Vienna, or

MEASURING THE SIZE OF SCIENCE
Table 7.1
Number of Scientific Authors by Country

World Total 126,055

USA	± 52,195	Mexico	152	Tanzania	10	
England	11,186	Greece	147	Madagascar	9	
(UK	13,103*)	Chile	113	Malta	9	
USSR	10,505	Nigeria	97	New Guinea	9	
Germany	8,398	Venezuela	82 ↑ 99%	Canary Islands	9	
France	6,862			Saudi Arabia	8	
Japan	5,202	Taiwan	72 ↓ 1%	Sierra Leona	8	
Canada	3,997	Pakistan	68	Vietnam	8	
India	2,882	British		Guatemala	7 ↑ 99.9%	
Italy	2,733	West Indies	63			
Australia	2,038	Lebanon	58	(West Africa)	7 ↓ 0.1%	
Switzerland	1,767	Turkey	58	Congo	6	
Czechoslovakia	1,718	Uganda	57	Cuba	6	
Sweden	1,650	Iran	52	French Morocco	6	
Netherlands	1,412 ↑ 90%	Portugal	51	Zambia	6	
		Malaysia	41	Burma	5	
Scotland	1,332* ↓ 10%	Singapore	38	Iceland	5	
Poland	1,305	China	36	Morocco	5	
Israel	1,125	Thailand	34	CA	4	
Hungary	1,039	Uruguay	34	Monaco	4	
Belgium	924	Kenya	34	Senegal	4	
Denmark	728	Iraq	32	Liechtenstein	4	
Austria	646	Peru	32	British Honduras	3	
Rumania	557	Philippines	32	Libya	3	
Finland	447	Hong Kong	30	Tunisia	3	
Norway	432	Sudan	28	Afghanistan	2	
Wales	384*	East Africa	27	Borneo	2	
Bulgaria	376	Rhodesia	25	Cyprus	2	
South Africa	338	Ghana	24	Ecuador	2	
Argentina	299	Algeria	22	Fiji	2	
United		Korea	22	Kuwait	2	
Arab Republic	293	French		Luxemborg	2	
Yugoslavia	288	West Africa	16	Northern Sudan	2	
Spain	277	Indonesia	14	New Caledonia	2	
New Zealand	253	[Africa]	13	Syria	2	
Brazil	206	Ceylon	11	Virgin Islands	2	
Northern Ireland	201*	Ethiopia	11	Once only mentions	33	
Ireland	156	Costa Rica	10			

Table 7.2
Great Scientific Cities of the World and of the United States
(number of scientific first authors in 1967)

World		World		United States		United States	
Moscow	4,982	Orsay	350	New York	2,783	Murray Hill, N.J.	366
London	2,915	Sydney	341	Washington	1,506	Denver	350
Paris	1,804	Hamburg	331	Boston	1,453	Oak Ridge	327
Tokyo	1,681	Basel	323	Philadelphia	1,407	Detroit	325
Leningrad	1,309 ↑ 10%	Sofia	321	Chicago	1,404	Dallas	323 ↑ 20%
		Frankfurt	307	Los Angeles	1,205		
Prague	882	Geneva	306	Cambridge, Mass.	1,010	Boulder	317
Kiev	728	Vancouver	296	Bethesda	911	Gainsville	311
Cambridge	720	Nagoya	293	Berkeley	869 ↑ 10%	Rochester, Minn.	307
Osaka	719	Melbourne	292			Bronx	298
Berlin	692	Brussels	290			Atlanta	296
Budapest	667	Uppsala	289	Pittsburgh	720	Davis, Calif.	294
Oxford	635	Amsterdam	285	Madison	702	Cincinatti	290
Munich	627	Delhi	283	Ann Arbor	679	Argonne	286
Ottawa	571	Freiburg	280	Seattle	648	Iowa City	282
Stockholm	546	Calcutta	276	Cleveland	632	Lafayette	281
Toronto	526	Sheffield	276	San Francisco	613	Ames	256
Montreal	522	Liverpool	275	Houston	576	Austin	253
Milan	480	Gothenburg	274	Minneapolis	575	Syracuse	252 ↑ 28,974
Warsaw	460	Kharkov	270	New Haven	543		
Copenhagen	454	Bristol	264	St. Louis	537	University Park, Pa.	249
Zurich	444	Helsinki	263	Princeton	480	East Lansing	247
Birmingham	427	Strasbourg	259	Buffalo	473	San Diego	246
Kyoto	418	Rehovoth	252	Columbus	459	Upton, N.Y.	246
Glasgow	386	Sendai	251	Urbana	458	La Jolla	231
Rome	386	Oslo	250 ↑ 33,020	Palo Alto	451	Indianapolis	227
Manchester	386			Rochester	448	Dayton	208
Jerusalem	374			Pasadena	444	Salt Lake City	203
Bucharest	370 ↑ 20%	Leeds	249	Stanford	406	New Brunswick, N.J.	197
Vienna	364	Gif-Yvette	248	Ithaca	405	Nashville, Tenn.	194
		Edmonton	243	Durham, N.C.	394	Lexington, Ky.	194
Edinburgh	357	Minsk	231	New Orleans	369	Portland, Oreg.	190

50% World Pop. of Sci.

Edinburgh, or, to put it another way, that single city has as many scientific authors as Bulgaria or South Africa. Rehovoth has as many as New Zealand, and Haifa has more than Mexico or Greece. Thanks to these figures, one can see for the first time the way in which each country divides its scientific strength among its principal cities (table 7.3). The distribution is very different from that of the nonscientific population in these cities. For what it is

Table 7.3
Percent Distribution of Publishing Scientists in Major Countries
by Chief Cities

United States		Soviet Union		United Kingdom	
New York	5	Moscow	50	London	23
Washington, D.C.	3	Leningrad	13	Cambridge	5
Boston	3	Kiev	7	Oxford	4
		Kharkov	5	Birmingham	3
				Glasgow	3
				Manchester	3
France		West Germany		Japan	
Paris	26	Berlin	8	Tokyo	33
Orsay	5	Munich	7	Osaka	14
Gif-Yvette	3	Hamburg	4	Kyoto	8
				Nagoya	6
Canada		Italy			
Ottawa	14	Milan	18		
Toronto	13	Rome	14		
Montreal	13				
Czechoslovakia		Hungary		Poland	
Prague	52	Budapest	66	Warsaw	35
Sweden		Denmark		Switzerland	
Stockholm	33	Copenhagen	62	Zurich	25

This list includes all the scientific cities of the world outside the United States down to the size of Jerusalem, and gives the percentage contributed by each of those cities to the scientific publishing manpower of each respective country.

worth, one may note (table 7.4) that Israel's scientific work appears to be distributed in a mode rather similar to Japan, and quite unlike the highly centralized countries, like the Soviet Union, Czechoslovakia, Hungary, and Denmark, and on the one hand, and the highly decentralized United States, Germany, or Italy, on the other hand. One may note, also, as a matter of interest, that it is possible on this basis to pinpoint for every country all the major cities of scientific activity, including those where one guesses that a considerable amount of the scientific work must be other than that published openly.

Probably the greatest interest in this new data arises from an investigation of the way in which the smaller countries compare in their shares of the world's science as against their shares of the wealth and of the population. Again, for these small countries as for the big, it is evident enough that it is the share of the wealth that determines that of the science (see fig. 7.1). The outstanding positive deviation, not approached by any other country so far as I know, seems to be the case of Israel. This country (see table 7.5) has about 0.15% of the world's GNP, which is about twice what

Table 7.4
Distribution of Publishing Scientists in Israel by City

	Number	Percent	Chief Center
Jerusalem	374	34	Hebrew University
Rehovoth	252	23	Weitzmann Institute
Haifa	177	16	Technion
Tel Aviv	93	8	University
Yavne	38	3	Atomic Energy
Nes Ziona	27	2	Institute of Biological Research
Petah Tiqwa	27	2	Medical Research
Be'er Sheva	24	2	Negev Research
Tel Hashomer	20	2	Government Hospital
Ramat Gan	15	1	Bar-Ilan University
All else	78	7	
Total	1,125	100	

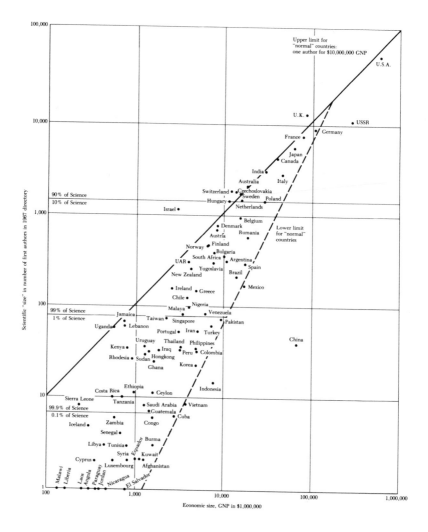

Figure 7.1. Figure does not include 24 small countries (GNP from 50 to 1800 million dollars) totaling 0.5% world GNP but with no scientists on directory in 1967.

Table 7.5
Percentage Shares of the World's Population, GNP, and Scientific
Publishing Manpower

	Population	GNP	Scientists
United States	5.9	32.8	41.5
Soviet Union	7.0	15.6	8.0
United Kingdom	1.6	4.8	8.1
France	1.4	4.5	5.4
Japan	2.9	3.6	4.1
Italy	1.5	2.6	2.2
Canada	0.6	2.2	3.2
India	14.4	2.2	2.3
Switzerland	0.2	0.6	1.4
Israel	0.08	0.15	0.9
Rest of Near East	2.5	0.85	0.4
Latin America	7.0	3.7	0.9

it would have if the wealth were equally shared amongst all the earth's peoples. Israel is, therefore, rather richer than the norm; it is far less fortunate than the United States or Canada or France, rather less than Switzerland or the United Kingdom, similar to the Soviet Union or Italy, better off than Japan, and enormously better off in wealth per capita than other countries in the Near East and all nations in Latin America.

The per capita wealth, however, only determines to some extent the per capita scientific strength. As is seen from the figure, the total scientific strength is well correlated with the economic wealth for most of the countries of the world. The most scientifically developed nations, from the biggest to the smallest, all cluster along a line which corresponds to one author on the International Index for every ten million dollars of GNP. It must be remembered that this number of authors is only a relative, and not an absolute, index of the gross number of scientific workers. We may normalize the data to some extent by noting that in 1964, or thereabouts, most of the advanced nations, big and small, were spending about 1 percent of their GNP on scientific research (basic and applied, but *not* including any of the considerable expenditures in development research).[6] If we were to assume, just as a talking point, that the International Index was listing for each country only about one-quarter of all those scientists who are doing research, and that the other three-quarters are doing work that will not be published this year (or perhaps at all!) then it will follow that each country is spending about 10^5 for every four scientists—an amount which gives an expenditure of $25,000 per scientist to cover his salary and his expenses in equipment and overhead.

Only the better-known scientific nations reach anywhere near this limit; most of the lesser nations fall below it, and the smaller the nation the more it can fall below the standard. China is, of course, a special case, since the openly published scientific literature is vanishingly small. The evidence would show, by the way, that if China published openly in keeping with its accepted present economic size, it would be a country of the same scientific magnitude as Japan, certainly not very much more or less. The manpower figures we derive for the Soviet Union are also much lower than we expect on other evidence, chiefly because of the "first author" phenomenon discussed previously, where many names are subsumed by that of the institute director.

The other very low countries include those cases where the state has more or less purposely a nonscientific policy, giving priority to other political or economic or military purposes: Indonesia, Cuba, Vietnam, Korea. It is a little surprising that Pakistan and Mexico are among this group, too, and it is equally surprising that there seems to exist some general minimum of science that is difficult to transgress so that there is a rather well-defined lower limit for scientific size versus economic size of the group in the case of all nations.

At the other end of the scale, above the line marking the relation between scientific and economic size for the most sophisticated nations, there is only one single case: that of Israel. This country has a scientific population, according to this indicator, which is more than a factor of 3 higher than that which one would expect by the standards even of such highly developed scientific nations as Britain, Switzerland, Hungary, and Czechoslovakia. From the fragmentary comparable data which exist for manpower, one can be reasonably sure that the indicator we have used is not grossly in error for the state of science in Israel. By any approximate standards of comparison, Israel has about two-thirds of the scientific manpower of Czechoslovakia in agreement with the above figure. I do not believe that the statistics are to be trusted with any higher confidence than half an order of magnitude, but the indication here is unmistakable; Israel has a scientific research population much larger relative to its size than any other nation in the world.

The special finding for science in Israel has clear historical roots in the special capacities of the Jews for such work and their motivation to it as "the people of the book" (compare, for example, the similarly high proportion of Jews among scientists in the United States, the Soviet Union, etc.) as well as in the great ease with which scientists have been able to emigrate from other countries taking their most portable of trades with them. It is not so difficult to find good reasons for explaining why and how Israel should be in this position. What to do with this position, now that it has been attained, is quite another matter. It seem obvious that Israel must pay special attention to the devlopment of science-based industries that can bring this abnormally large population of scientists to the special level of productivity it might achieve. The maximum use of sophisticated science in military service is another obvious question, though I do not pretend to know any answers. The most obvious thing is that one must treat scientific research as a considerable export industry, doing jobs for other nations. Unless good progress is made here, it will, at the least, be an oddity that any nation can afford to be so wise.

8

Citation Measures of Hard Science, Soft Science, Technology, and Nonscience

Perhaps the fundamental problem of those who work in the scientific information industry is that it is not just that special part of an information industry that happens to deal with material having a scientific content. Technical librarianship involves much more than librarianship applied to books with an esoteric vocabulary and much mathematics. It is somewhat like the dilemma of the man who tried to write a book on Chinese medicine by first reading one on China and then another on medicine and then "combining his knowledge." Scientific literature differs in a greater degree from the rest of literature, even from the rest of scholarship, than Chinese medicine differs from the rest of medicine. Moreover, I suggest, the literature of the scientist differs essentially in its social role from that of the nonscientific scholar—it is not just a matter of a different substantive content.

My reason for choosing this problem as a contribution for such a strategic conference is that I know well as a historian of science that the greatest and most useful advances in our technologies have not come from the applied research of trained people trying to make themselves useful to society, but rather from basic research aimed at furthering understanding and curiosity, and powered by the latest instrumentation that the useful people have devised. I suspect that all the new indexing tools and computer handlings will be more useful to basic research in understanding scientists than they will to solving practical problems for which they are designed. It seems clear, however, that it is only such new understanding that can bring success, whatever solutions ultimately emerge.

I take, therefore, as my problem here the elucidation of what it is apart from substantive content that distinguishes scientific information from other information. It must be noted that such research is only possible because quite new and important unobtrusive measures of scientific literature have become available as a sort of waste product of the large-scale commercial operation of the *Science Citation Index* as a radically new tool for practical needs of scientists.[1] Such computerized bibliographies give immediately counts that surpass in accuracy and scope all previous data, sometimes by orders of magnitude. The problem has been to develop counts of such things that might be a diagnostic of whether a piece of scholarship or a field of knowledge acted as "science" or as "nonscience."

A now classical paper by Deutsch worked out in some detail the implications of a suggestion by Conant that the essential difference between the two modes of scholarship was that of "cumulation" versus "noncumulation."[2] It was seen that cumulation in this sense implies not merely growth, nor indeed growth at compound interest, but rather the existence of a tightly integrated structure for the sciences. Evidently the prototypes of the other side, identified as "the humanities," grew (perhaps almost as fast), contained specialties and fashions just as science, but had something different from the integrated structure of cumulation.

Some empirical evidence was fed into this qualtitative analysis by Storer[3] who also broke new ground in setting up an analysis of "hard science" versus "soft science" and by implication of "nonscience." Perhaps the most useful thing is that he got us over the barrier that nobody likes to be "soft." Storer pointed out that it was not a distinction of "hard" in the sense of "difficult" as against subjects that were soft and easy. The words were being used in such ways as brittle, unyielding, impersonal for the hard, and gentle and adapting for the soft. Perhaps the usage in "hard and soft sell" in advertising is nearer to that we want to agree with our intuitive senses. At all events, it was shown that the intuition could be related quite simply to scales provided by the amount of tabular presentation and the amount of usage of full given names or initials only in the references in the literature of these fields.

The first of these is obviously a valid though rough measure of the use of quantitative methods in the substantive content, which is fair enough for those subjects where one might not otherwise know in advance if the field were wordy or mathematical. The other measure is a horse of a different color. It is a measure of impersonality in the relations between authors and leads to quite new speculations. A high degree of given name personality would indicate an invisible college structure of authors known to each other as warm bodies rather than as labels on literature. Though one might suppose that this latter measure was largely a matter of fashion and unthinking tradition, the index it generates seems rather consistent and accurate enough to show a clear time variation toward impersonality in the social sciences.

Another measure that might be used, but has not been, to distinguish science from nonscience, and to sort the soft from the hard, would be something that reflected on the social place of each branch of learning in society. The whole life of a discipline within a university is determined by whether it is used for education or for training. In some fields, such as history and philosophy, most of the embryonic researchers get their Ph.D.'s and then proceed toward some sort of career as a teacher. In that case society is paying for students to become teachers to beget students; research becomes an epiphenomenon. In the most "scientific" departments at our universities only about 20 percent of the Ph.D. output is fed back into education, and society gets for its investment not only the education of the undergraduate young and a by-product of research, but also the training of Ph.D.'s who become employed in the nonuniversity world. If we take this measure of feedback, it provides a functional spectrum showing the range from the hard sciences to the soft sciences and thence to the humanities.[4] Further work here showed that universities in the United States had an overall feedback rate of about 50 percent and that it took about 100 teachers with doctorate qualification for every 15 new doctorates produced per year.[5] This input-output model of undergraduate and graduate education surely deserves further elaboration on a field-by-field basis.

Another series of such social parameters has been demonstrated

Table 8.1
Percent of Ph.D. Graduates Employed in College or University

Chemistry	23.7
Engineering	25.1
Psychology	27.3
Physics	38.7
Microbiology	56.5
Botany	63.4
Mathematics	68.7
Zoology	70.2
Other Bioscience	73.5
Economics	75.4
Sociology	83.2
History	88.8
English	91.2
Foreign Languages	91.9
Political Science	93.9

Source: Adapted from Kenneth M. Wilson, "Of Time and the Doctorate," *Southern Regional Education Board Research Monograph No. 9* (Atlanta: 1965).

by Warren Hagstrom in his classical study of the scientific community.[6] In these he shows, for example, that the scale from hardest sciences to softest humanities coincides in hierarchy with the decreasing scale of agreement with the proposition that "major professors often exploit doctoral candidates"—a clear measure of collaboration through economic subservience. Perhaps less emotionally loaded are his findings that there are different spectrums governing the amounts of publication in research articles, in review articles, and in books. The characteristic differences from field to field seem to form a very clear pattern, depending only on one or two independent social parameters.

At this point it becomes evident that we cannot and should not artificially separate the matter of substantive content from that of social behavior. In order to deal with quantitative, highly orderly, rather certain findings, a special sort of social relation between participants is called for. This relationship changes with time; we note, for example, that in Galileo's day it was an embarrassment to find books which were not old classics but authored by scholars that were still alive; by the scientific revolution of the mid–seventeenth century, the pace increased to the point where scholars met as a learned academy and where consequently the scientific paper and journal could be invented. The time span is ever decreasing and changing our behavior. It does this because, essentially, scholarship is a conspiracy to pool the capabilities of many men, and science is an even more radical conspiracy that structures this pooling so that the totality of this sort of knowledge can grow more rapidly than any individual can move by himself. The humanities, by resting with the capability of the individual, eschew this growth rate and certainty. As we find ways of being certain through impersonality and mathematics, so the humanities are left with fewer of the problems. It is in this way that natural philosophy was transformed into science, and, in general, it is in this way that the different substantive contents have erected different social apparatuses of information pooling and exchange.

I turn immediately from the humanistic philosophy of this process to a more certain, or at least a more quantitative, formulation of the matter in hand. A scholarly publication is not a piece of informtion but an expression of the state of a scholar or group of scholars at a particular time. We do not, contrary to superstition, publish a fact, a theory, or a finding, but some complex of these. A scientific paper is at the same time more and less than a concept or a datum or a hypothesis. If the paper is an expression of a person or several persons working at the research front, we can tell something about the relations among the people from the papers themselves.

It is in this sense that several of us have used bibliographical references and collaborative authorships as indications of social links.[7] Though collaborative authorship has proved useful as a means of analyzing invisible colleges and in-groups,[8] it is perhaps not so useful as a means of differentiating between one sort of scholarly field and another. Hirsch and Singleton,[9] in a paper that has lain unpublished for too long, demonstrated quite effectively that the amount of collaborative authorship in a field was directly proportional to the amount of economic support received by the workers. My own sociological analyses indicate also that collaboration arises more from economic than from intellectual dependence and that the effect is often that of squeezing full papers out of people who only have fractional papers in them at that particular time. At all events, the amount of collaborative authorship measures no more than the economic value accorded to each field by society. A soft subject highly subsidized would become as collaborative as high-energy physics; it would be interesting to see what happens to urbanology.

The sheer amount of bibliographical references provides a handy measure of social linkage. The etymology of "scholarship" indicates that it derives from the *scholia,* the added explanatory footnotes put into school texts, so perhaps it is reasonable to identify the amount of such footnotage and referencing with our intuitive idea of "scholarliness." I would not like to push this point too hard by claiming, for example, that all papers with a dozen bibliographical references were more scholarly than all those with only 10. Nor would I dream of maintaining that all papers with 10 references were of similar scholarliness. I would, however, direct your attention to three relevant questions.

1.   Why is the norm of scholarship a paper with ca. 10 to 22 references?

2.   What sort of paper lies far above this norm?

3.   What sort of paper lies far below it?

I cannot answer the first question except to remark that for a literature growing as ours does at about 7 percent per annum the number of references back is then of the same order of magnitude as the corpus of past literature. Thus, each old paper gives rise roughly and on the average to about 1 citation per year. This seems a sort of natural rate of procreation, and one might have a very different world if each paper gave rise to, say 10 citations per year. There seems to be a slow but steady increase in referencing in all fields. Even *Physical Review* contained few explicit references before World War I. Parker et al. find for the social sciences that references per article jumped from 8.4 and 9.4 in 1950 and 1955 to 15.2 and 15.5 in 1960 and 1965.

As to the second question of very scholarly papers, we must find it a little disappointing for quite clearly the papers with large numbers of references are not creative scholarship at all, but rather they are review papers condensing and giving access to a pile of recent literature in some particular specialty. It seems, from a first inspection, that this is just as likely to happen in art, history, or comparative literature as in international law, or even in the more familiarly hard sciences areas of physics and biochemistry. The only new finding I have to offer is that there seems much less distinction than I had expected between these review papers ánd the research papers; they merge into each other insensibly.

The third question is a bit more productive; unscholarly papers with hardly any explicit references certainly exist in profusion, even in decent academic subjects like economics and mathematics. Trivially and quite typically such unreferenced papers occur when an experienced scientist or librarian makes an *ex cathedra* pronouncement out of his innate knowledge of what should be or what is. The same is true in all fields of knowledge; there is perhaps a place for pronouncements, but it must be firmly stated that there is an implication that the culminating scholarship does not exist, exists but is irrelevant, or exists relevantly but is unknown. Scholarliness as I have defined it may be taken not just as a diagnostic but also as prescriptive for a cumulating knowledge system. Not so trivially, it is found that a very large class of technical news magazines is largely unreferenced in this way. I suggest that science may be *papyrocentric* but that a large part of technology is *papyrophobic*. If you want to make capital out of technological discovery, the last thing you want is that open publication that determines intellectual private property for the sciences. Why then do we have technical magazines of this sort? I must suggest that they function as a news medium outside the knowledge system. Perhaps, more likely, it is not the text, but rather the advertisements that are the most prized part of the package in this literature. We know that when an engineer moves, it is his catalogs rather than his journals that go with him.

If indeed some parts of technology are low in scholarliness, it makes one wonder if direct economic value is an inhibitor of information flow. Is it prescriptive in the sense that technologists do not use references and, therefore, do not need to be fed any papers which might be embodied in their own contributions? I doubt this. The technologists seem to want papers all the more, and the resultant conflict or paradox is probably at the heart of any so-called literature crisis or information problem. It is perhaps worth noting that citation among patents is also amazingly light so that there are no "highly cited" patents, and the system has only fragmented short chains of self-citation rather than the diversified branching network found in papers and in scientists. Also worth noting is the fact that this all seems little different in the Soviet literature in spite of the different economic basis of technology. It seems that technologists differ markedly from both scientific and nonscientific scholars. They have a quite different scheme of social relationship, are differently motivated, and display different personality traits.

The amount of referencing by itself is certainly not enough to give us the distinctions we have been looking for between science and nonscience, hard and soft. What is needed is some measure of the texture of this system of referencing and citation, whether it be applied heavily or lightly. We need not only the rate and density of the procreative system some but knowledge of its metabolism or its eugenics.

In my first (and, I am afraid, grossly overcondensed) paper on scientific networks,[10] I was able to use the *Science Citation Index* data for millions of papers to show conclusively that there was not a single population of references but, rather, two overlapping populations. On the one hand, there was a fairly uniform raiding of the archive of all the available literature, past and present, with only a slow secular decrease in the usefulness of literature as a function of its age. On the other hand, there was something which

I called an "Immediacy Effect," a special hyperactivity of the rather recent literature which was still, so to speak, at the research front. I want to stress that this immediacy is something quite different from the normal aging of the literature, for this has been misunderstood by at least one commentator.[11]

The normal aging is not so simple either. For literature more than about fifteen or twenty years old, the time spread of references matches rather well the known growth of the literature. A modern work will refer to almost twice as many papers from 1913 as from 1900 because there were twice as many papers published in 1913 as there were in 1900. The number of citations per paper published does fall off with time, but probably only with a halving in value every twenty or thirty years as the knowledge gets packed down and outmoded. The near constancy of every paper ever published producing about 1 citation per year on the average is a bit misleading. One must not forget that the number of citing papers is growing exponentially, too, and doubling every ten years or so. Thus the number of citations per *citing* paper is actually halving every ten years, and it is only the near equality between the growth rate of literature and the half-life of obsolescence that gives one a near constant balance just noted, of 1 citation per archival paper per year.

The phenomenon I invoke as an immediacy is a much increased use of the last few years of papers over and above the natural growth of the literature and its normal slow aging. The present year and last year of accrued literature does not show up so strongly because it takes a while for the information to become well known and available, and there is a time lapse, now well documented, between the communication and its publication in the journals embodied as a reference. Papers from about two years ago may be cited six times as much as the normal archival rate, from three years ago five times, and from nine years ago three times. These rates represent averages over the whole field of science and may be faster or slower field to field. The effect of this extra immediacy of interest in recent work is that the references fall off with relative age (i. e., time difference between citing and cited paper) very much more rapidly than the normal 13.5 year halving time given by the *Science Citation Index* statistics. Over the SCI literature the rate for recent references is a halving in number with every five years of age of reference. MacRae finds this figure for sociology and the soft sciences, but an even steeper immediacy effect of a halving in three years for physics and the biomedical sciences.[12] Stoddart found the half-life of citations in geography decreased from an archival sixteen years in 1954 to a steep research front of four years in 1967, the change being quite regular and marked.[13]

I think it is, however, of relatively little value to determine the exact rates of decline in recent references as has often been done in the determination of the half-life of references.[14] What is more important than the exact profile of the falloff in immediacy effect is the proportion between this research front immediacy, on the one hand, and the normal archival use of the literature, on the other hand. Meadows has suggested an immediacy index which is the ratio between the numbers of references to the last six years compared with those to work more than twenty years old, but I find little payoff in the additional complexity.[15] It seems to me quite reasonable to take as a valid measure the proportion of the references that are to the last five years of literature. The reason I choose five years is that ten is too much and three too few. Ten would give a 50 percent increase from exponential growth, even without any immediacy effect. Three years does not give a long enough time for consistent figures, since the first two years contribute little (the paper is not well disseminated), and there is a fluctuation caused by the cycle of the calender year and periodic annual publication. Several previous workers have used this instinctively, and the choice seems felicitous.

To get an estimate of the range of values of this parameter, we may note that a literature growing at a rate of 5 percent per annum doubles in size in 13.9 years and contains about 22 percent of all that has been published in its last five years of publication. A field growing at the most rapid rate experienced of 10 percent per annum, with a doubling time of 6.9 years, has within the last five years about 39 percent of all its archive. Price's Index (as I might call it) will vary from 22 percent for normal growth to 39 percent for most rapid growth for a field that is purely archival, raiding all the literature that has gone before equally, with only a gradual secular decline with aging, and without this special immediacy of an active research front (see fig. 8.1).

It is difficult to estimate exactly what Price's Index would be for a subject that was all research front and no general archive, but I hazard a guess that 75 to 80 percent is the range to look for, again depending on rate of growth. Any mixture of the two types of referencing will result in an intermediate value of the index, and we, therefore, derive a useful indication of the balance as it varies from subject to subject. Parenthetically, I would remark that there is nothing here to restrict this Price's Index or the number of references per paper to being used over a whole field. It can be given micro- as well as macrouse, as a means of evaluating (with what uncertainty we know not) a journal, an institution, the idiosyncrasies of even a single person, or just a single paper. It goes without saying that this paper in its published form will have to have about the norm of 16 references at least, and I would suppose that however I make my choices in a field like this, all but 2 or 3 references will be to work published here after 1963.

My study of the *Science Citation Index* statistics showed that the average of all fields covered was a Price's Index of just over

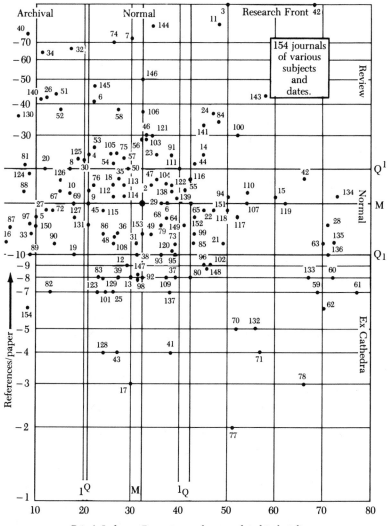

Price's Index = Percentage references dated in last five years

Figure 8.1. Price's Index

Table 8.2
Summary of Data Collected from 162 Journals
of Various Subjects and Dates

	Lower Quartile	Median	Upper Quartile
Number of references per article	10	16	22
Percent references in last five years	21	32	42

Data from *Science Citation Index* statistics:

	1964	1965	1966	1967	1968
Citation from source journal items per source journal item	11.77	12.39	11.26	11.14	12.00
Percent total citations in last five years	55.13	55.30	52.56	52.49	52.11

50 percent, and from this I deduce that about half of the references go to an even spread over the entire archive while the other half go to the research front. For the archive this means that for every hundred papers in the archive we have about 7 new papers each year with their 100 references back of which about 50 are citations to the archive. It follows that the chance of an archival paper being cited in any particular year is about one-half so that roughly half of the citable papers lie uncited in any year (actually it seems to be about 40 percent uncited, 60 percent cited). For the research front which contains about 30 papers within the last five years, we have another 50 citations available, and this results in an average of about 1.67 citations per paper cited—again in agreement with SCI statistics.

Perhaps the most important finding I have to offer is that the hierarchy of Price's Index seems to correspond very well with what we intuit as hard science, soft science, an nonscience as we descend the scale. At the top, with indexes of 60 to 70 percent we have journals of physics and biochemistry; a little lower there

Table 8.3
Percent of Citations Dated in Previous Five Years

American Behaviorial Scientist	1965	43.6
American Documentation	1955	44.2
	1960	47.0
	1965	59.8
American Educational Research Journal	1965	36.0
American Psychologist	1950	55.8
	1955	51.0
	1960	51.4
	1965	47.6
American Sociological Review	1950	38.8
	1955	47.4
	1960	43.2
	1965	35.2
Audio-Visual Communication Review	1960	45.6
	1965	44.2
Behavioral Science	1960	48.8
	1965	49.4
Journal of Abnormal and Social Psychology	1950	44.6
	1955	48.0
	1960	40.0
	1965	39.2
Journal of Advertising Research	1965	41.6
Journal of Broadcasting	1960	31.4
	1965	38.0
Journal of Communication	1965	36.0
Journal of Educational Psychology	1950	31.0
	1955	35.4
	1960	37.8
	1965	51.8
Journal of Verbal Learning and Verbal Behavior	1965	47.4
Journalism Quarterly	1950	38.4
	1955	47.0
	1960	39.2
	1965	42.4
Psychological Bulletin	1950	35.0
	1955	44.8
	1960	34.0
	1965	37.8

Table 8.3 (continued)

Public Opinion Quarterly	1950	42.4
	1955	37.8
	1960	47.0
	1965	38.2
Sociometry	1950	46.8
	1955	44.2
	1960	32.4
	1965	37.4

Source: Adapted from Edwin B. Parker, William J. Paisley, and Roger Garrett, *Bibliographic Citations As Unobtrusive Measures of Scientific Communication* (Stanford: Stanford University Institute for Communication, October 1967).

are publications like *Radiology* (58 percent) and *American Journal of Roentgenology* (54 percent).[16] *American Sociological Review* stands at 46.5 percent, and a study by Parker et al. shows most of the other social sciences clustering around 41.9 percent ± 1.2 percent,[17] while a pilot project investigation of my own covering 154 journals of various brands of scholarship[18] (see table 8.4) shows that the median over all fields of science and nonscience was 32 percent with quartiles at 21 percent and 42 percent. I find some slight indication of a slow increase of the index with time for any particular field, but the average over the whole *Science Citation Index* has fallen with time as more of the softer sciences have been included.

The journals with an index higher than the upper quartile (i.e., above ca. 43 percent) are unquestionably of the hard science variety, and those with an index higher than 60 percent are just the ones where competition, fashion, invisible colleges, and all direct social action symptoms are evident. At the other end of the

Table 8.4
Price's Index for a Variety of Journals

	Title	No. of Articles		Average no. of refs./ article	Percent of all refs. dated within last 5 yrs.
1	Acta Physiologica	12	1954	17	40
2	Acta Physiologica	11	current	19	34
3	Advances in the Physical Sciences (Uspekhi)	3	current	100	50
4	American Antiquity	8	recent	25	21
5	American Historical Review	15	1925	14	10
6	American Historical Review	6	1950	41	22
7	American Historical Review	3	current	73	30
8	American Journal of Archaeology	9	recent	22	17
9	American Journal of Botany	11	recent	18	21
10	American Journal of Economics and Sociology	8	recent	18	17
11	American Journal of International Law	3	recent	83	48
12	American Journal of Mathematics	12	recent	9	29
13	American Journal of Philology	16	recent	8	30
14	American Journal of Physicial Anthropology	8	recent	25	45
15	American Journal of Sociology	12	recent	17	60
16	American Literature	11	recent	11	4
17	American Mathematical Monthly	52	recent	3	30
18	American Naturalist	11	recent	19	26
19	American Philosophical Quarterly	10	recent	10	18
20	American Quarterly	8	recent	22	12
21	American Sociological Review	18	1936	11	49
22	American Sociological Review	15	1951	15	45
23	American Sociologial Review	8	recent	25	35

	Title	No. of Articles	Average no. of refs./ article	Percent of all refs. dated within last 5 yrs.
24	American Zoologist	6 recent	37	47
25	Anatomischer Anzeiger	30 1928	7	26
26	Anatomischer Anzeiger	6 1950	42	12
27	Anatomischer Anzeiger	13 current	15	12
28	Anesthesie Analgesie Reanimation	15 1957	13	71
29	Anesthesie Analgesie Reanimation	13 current	16	36
30	Anglican Theological Review	5 recent	24	20
31	Annales Medico Psychologiques	18 1930	11	31
32	Annales Medico Psychologiques	4 current	66	17
33	Annals And Magazine of Natural History	16 recent	12	9
34	Annals of the Association of American Geographers	3 recent	64	11
35	Annals of Botany	10 recent	20	27
36	Antiquity	19 recent	12	27
37	Architectural Science Review	15 recent	8	39
38	Archiv für Physikalische Therapie	19 1949/50	10	31
39	Archiv für Physikalische Therapie	24 current	8	27
40	Art Bulletin	3 recent	76	8
41	Art Journal	13 recent	4	38
42	Astrophysical Journal	10 recent	20	66
43	British Journal of Aesthetics	28 recent	4	27
44	British Journal of Experimental Pathology	13 recent	23	43
45	British Journal of Psychology	13 Batch I	16	21
46	British Journal of Psychology	7 Batch II	30	33
47	British Journal of Psychology	12 Batch III	20	35

	Title	No. of Articles		Average no. of refs./ article	Percent of all refs. dated within last 5 yrs.
48	British Journal of Psychology	17	Batch IV	12	27
49	British Journal of Psychology	26	recent	12	34
50	British Journal of Sociology	9	recent	22	29
51	Bulletin of the Geological Society of America	5	Batch I	44	14
52	Bulletin of the Geological Society of America	6	Batch II	38	15
53	Bulletin of the Geological Society of America	8	Batch III	27	22
54	Bulletin of the Geological Society of America	9	Batch IV	23	26
55	Bulletin of the Geological Society of America	13	Batch V	18	41
56	Bulletin of the Geological Society of America	5	Batch VI	39	32
57	Civil War History	15	1955	13	9
58	Civil War History	5	current	38	27
59	Communications of the ACM	27	Batch I	7	69
60	Communications of the ACM	27	Batch II	8	72
61	Communications of the ACM	30	Batch III	7	77
62	Communications of the ACM	35	Batch IV	6	70
63	Communications of the ACM	18	recent	11	70
64	Comprehensive Psychiatry	14	Batch I	15	39
65	Comprehensive Psychiatry	12	Batch II	16	42
66	Comprehensive Psychiatry	13	Batch III	16	37
67	Comprehensive Psychiatry	11	Batch IV	18	15
68	Comprehensive Psychiatry	15	Batch V	14	37
69	Comprehensive Psychiatry	8	recent	16	18
70	Computer Journal	19	recent	5	52
71	Comtes Rendus des Seances	51	current	4	57
72	Criticism	10	recent	15	13
73	Czechoslovak Journal of Physics	18	recent	11	39

	Title	No. of Articles		Average no. of refs./ article	Percent of all refs. dated within last 5 yrs.
74	Ecological Monographs	3	recent	70	26
75	Ecology	8	recent	24	26
76	Économic Botany	11	recent	18	21
77	Economic Journal	21	1907	2	51
78	Economic Journal	28	1925	3	66
79	Economic Journal	7	current	13	36
80	Electronic Computers	25	recent	8	42
81	English Literary History	9	recent	23	8
82	French Review	19	recent	7	13
83	Geographic Journal	21	recent	8	23
84	Geological Society of America Bulletin	8	recent	34	48
85	Geophysical Journal	17	recent	11	43
86	Geophysics	19	recent	12	24
87	German Review	12	recent	13	6
88	Isis	12	recent	18	8
89	Islamic Studies	8	recent	10	9
90	Journal of Aesthetics and Art Criticism	19	recent	11	14
91	Journal of American Academy of Religion	6	recent	25	38
92	Journal of the American Chemical Society	24	1920	8	32
93	Journal of the American Chemical Society	21	1950	10	36
94	Journal of the American Chemical Society	52	recent	17	50
95	Journal of American Statistical Association	20	recent	10	39
96	Journal of Analytical Chemistry	60	recent	9	45
97	Journal of Anatomy	16	recent	24	28
98	Journal of the Association for Computing Machinery	14	recent	8	31

	Title	No. of Articles		Average no. of refs./ article	Percent of all refs. dated within last 5 yrs.
99	Journal of the Atmospheric Sciences	15	recent	12	43
100	Journal of Biological Chemistry	7	recent	30	52
101	Journal of Cuneiform Studies	5	recent	7	24
102	Journal of Economic Entomology	22	recent	9	46
103	Journal of Economic History	8	recent	29	30
104	Journal of Experimental Botany	11	1950	19	36
105	Journal of Experimental Botany	8	current	24	25
106	Journal of the Geochemical Society	6	recent	37	33
107	Journal of Inorganic Chemistry	50	recent	16	54
108	Journal of the Institute of Wood Science	12	recent	11	26
109	Journal of Mathematical Analysis and Applications	27	recent	8	37
110	Journal Nuclear Physics	14	recent	17	54
111	Journal of Physical Chemistry	10	recent	21	40
112	Journal of Plasma and Thermonuclear Fusion	16	recent	17	26
113	Journal of Political Economy	11	Batch I	19	28
114	Journal of Political Economy	13	Batch II	17	38
115	Journal of Political Economy	13	Batch III	15	24
116	Journal of Political Economy	11	Batch IV	20	42
117	Journal of Political Economy	14	recent	14	52
118	Journal of Political Economy	14	Batch VI	15	50
119	Journal of Politics	12	1939	16	62
120	Journal of Politics	22	1950	10	39
121	Journal of Politics	8	recent	29	34
122	Journal Quantitative Spectroscopy and Radioactive Transfer	11	recent	19	38

	Title	No. of Articles		Average no. of refs./ article	Percent of all refs. dated within last 5 yrs.
123	Journal of Symbolic Logic	28	recent	7	22
124	Journal of Zoology	10	1949/50	21	8
125	Journal of Zoology	9	recent	24	19
126	Language	11	recent	19	15
127	Linguistics	11	recent	14	18
128	Mind	14	recent	4	24
129	Musical Quarterly	6	recent	8	22
130	Philological Quarterly	6	recent	36	7
131	Philosophy of Science	15	recent	13	21
132	Physical Review	47	1900	5	56
133	Physical Review	28	1925	8	67
134	Physical Review	13	1950	17	73
135	Physical Review	18	current	11	72
136	Physical Review Letters	15	recent	10	71
137	Physiological Reports	70	recent	7	38
138	Planetary and Space Science	11	recent	19	38
139	Plant Physiology	12	recent	17	40
140	Proceedings of the Geologists' Association	5	recent	41	12
141	Reviews of Modern Physics	7	Batch I	33	45
142	Reviews of Modern Physics	2	Batch II	101	68
143	Reviews of Modern Physics	7	recent	43	58
144	Reviews of Modern Physics	3	Batch IV	81	34
145	Reviews of Modern Physics	5	Batch V	47	22
146	Reviews of Modern Physics	4	Batch VI	50	32
147	Revue D'Assyriologie	6	recent	8	32
148	Semaine des Hopitaux	22	1950	9	45
149	Semaine des Hopitaux	19	current	13	41
150	Slavic and East European Journal	15	recent	13	10
151	Soviet Atomic Energy	13	current	15	47
152	Soviet Journal of Nuclear Physics	15	current	14	43

Title	No. of Articles	Average no. of refs./ article	Percent of all refs. dated within last 5 yrs.
153 Soviet Physics Crystallography	18 current	12	32
154 Studies in English Literature	27 recent	6	8

scale, I was most surprised to find that there are a considerable number of journals where the index is less than one would find with pure archive and no research front at all. *German Review, American Literature, Studies in English Literature,* and *Isis* are all samples of a very low index value, less than 10 percent. In a sense these fields find special reasons for citing older material which is indeed their universe of discourse. Among the sciences, I think one would find similarly that the taxonomic sciences, giving special place as they do to the first and earliest case examples, would also display an anomalous appearance of a negative amount of research front.

Such pathological cases apart, it would seem that this index provides a good diagnostic for the extent to which a subject is attempting, so to speak, to grow from the skin rather than from the body. With a low index one has a humanistic type of metabolism which the scholar has to digest all that has gone before, let it mature gently in the cellar of his wisdom, and then distill forth new words of wisdom about the same sorts of questions. In hard science the positiveness of the knowledge and its short term permanence enable one to move through the packed down past while still a student and then to emerge at the research front where interaction with one's peers is as important as the storehouse of conventional wisdom. The thinner the skin of science the more orderly and crystalline the growth and the more rapid the process. In essay 5 I guessed that the skin thickness of the research front might be put at something like fifty papers. After that number of contributions has been laid down, some sort of packing down production of a review paper or summary seems to be necessary, but perhaps that, too, varies from field to field.

It is also worth pointing out that the data for the population of journals studied show a scatter over the entire range of values of scholarliness and Price's Index. The two measures are therefore uncorrelated and quite independent parameters characterizing the journal in question.

I have already said that I regard the value of this work as being not only diagnostic, but also prescriptive, so let us look in closing at what suggestions it makes for the technology of scientific information. At the personal level of the scientific writer and the editor and publisher of journals, it tells that one might well be as careful about references as we are about titles, authorships, and proper presentation of data. (One must hope the effect is something more than a self-fulfilling prophecy!) For a research paper it should be exceptional to allow an author to eschew interaction with colleagues by citing too few references, and if he cited too many, perhaps he is writing a review paper rather than a research contribution. Similarly, if you want to make the field firm and tight and hard and crystalline you have to play with your peers and keep on the ball by citing their recent work.

At the systems level, perhaps these measures give prescriptions for the properties needed in an information service. If the people using the service use very few references, maybe they do not want scientific papers at all. If they give very many, they are striving for good old-fashioned scholarly completeness. If they write papers with a low Price's Index you just have to maintain an archival library as libraries have always been maintained and try all new tricks with indexes and computers. If, however, they write with a high Price's Index, it is some indication that the information system is most active at the research front and what we have is a current awareness system of interaction with peers, not with subjects or methodologies. For such people it seems obvious that citation indexing and the social engineering of invisible colleges produce results vastly superior to any possibilities of normal library or information organization.

In short, hard science, soft science, technology, and nonscience may be all different social systems, and each system must have its own special machinery for handling the processes of publication and communication among people at the research fronts and behind those fronts, too. I believe that the elucidation of problems like this provide the most exciting fundamental material for social science research, and that a proper understanding of science as a social system will wipe away a lot of naive misunderstanding which shrouds the business of science information and makes us hope for the wrong sort of expensive solutions to what seem to be the problems.

9

Some Statistical Results for the Numbers of Authors in the States of the United States and the Nations of the World

METHODOLOGY

The geographical breakdown of the world list of scientific authors provided by WIPIS provides one of the best unobtrusive indicators of science and technology research activity. Direct national statistics derived on a census or fiscal basis have the disadvantage that slight changes of definition due to the different socioeconomic conditions around the world may lead to large variations in absolute numbers of persons in any category of alleged "scientific and technical manpower." Counts obtained by allocating journals to their countries of publications lead to obvious peculiarities because of the significant number of high-status journals that are effectively international rather than national. Lastly, counts of the literature rather than the authors suffer from the considerable disadvantage that rates of productivity differ from field to field thus giving undue weight to fields where the norm is a short paper rather than an extensive monograph.

What do such numbers of authors indicate? I feel it is clear from the evidence that they give a rather effective universal measure of activity within the state or country directed toward all

---

Coauthored by Suha Gürsey.

that research work where the chief product is a publication of the type covered by the *Current Contents* journals which are the basis for WIPIS. The indicator covers, therefore, all fields of scientific and technical publication which are part of the world network of knowledge rather than of purely domestic interest. The number of authors publishing in a given year is by no means identical with the community of all authors who ever have published. Year by year the lists for each country will contain a different selection of names, some continuing for a long period, others for just a few years or very frequently for a single index. From a study of the publishing patterns I hazard a guess that the number of authors that have published is about four times as large as the number actually publishing in any year; this indicator may therefore be taken as showing the size of the research community.

With an accumulation now available of seven years of this directory it is possible to determine the relative sizes of the various nations and states with much greater confidence than previously. Moreover, by taking the regression coefficients of the relative sizes (or rather of their logarithms) it is now possible to give for the first time the relative rates of change of the entities involved. Logarithms are used throughout these presentations since it is these which are physically meaningful and follow normal distributions. It should be noted that the absolute numbers of authors and the absolute growths must be only partly due to increases in the manpowers; an undetermined but large part of each must be due to the gradually increasing scope of *Current Contents* which is, so to speak, expanding into the universe of journals. For this reason we shall present only comparative and relative values over the seven-year period. The data are presented here only in summary and in graphic form, partly to show the validity of the method, partly to enable others to test the indications against other evidence from other indicators.

Turning first to the distribution of publishing scientists in the fifty states of the United States, we present the logarithm of the number of authors (geometric mean over seven years) plotted in figure 9.1 against the manpower data taken from the *National Register of Scientific and Technical Personnel* for the year 1970 (NSF publication).

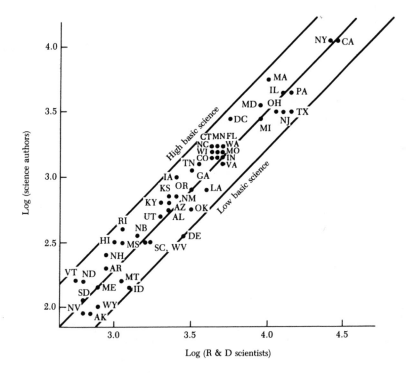

Figure 9.1.

Table 9.1

Geometric mean of number of authors and *National Register of Scientific and Technical Personnel* scientists for states, ranked by authors

State	Authors	R&D Scientists
California	11,024	28,462
New York	10,699	25,067
Massachusetts	5352	10,329
Illinois	4691	12,803
Pennsylvania	4661	14,005
Maryland	3719	9288
Ohio	3230	11,168
Texas	3206	13,749
New Jersey	3075	12,807
District of Columbia	2863	5897
Michigan	2833	8553
Wisconsin	1737	4691
Connecticut	1660	4447
Minnesota	1635	4411
Florida	1609	4566
Washington	1544	4767
Missouri	1534	4710
North Carolina	1506	3989
Indiana	1480	5396
Colorado	1459	5037
Tennessee	1304	3635
Virginia	1190	5300
Georgia	1176	3190
Iowa	1009	2588
Louisiana	840	3913
Oregon	790	3075
New Mexico	685	2485
Kansas	670	2268
Arizona	649	2203
Kentucky	627	1958
Alabama	583	2232
Oklahoma	577	3260
Utah	515	1941
Rhode Island	380	1097
Nebraska	348	1422
Delaware	340	2784
Hawaii	334	987

Table 9.1 *Continued*

State	Authors	R&D Scientists
South Carolina	326	1587
Mississippi	323	1166
West Virginia	303	1568
New Hampshire	254	853
Arkansas	196	892
North Dakota	167	636
Montana	162	1159
Vermont	161	540
Maine	144	757
Idaho	134	1191
South Dakota	113	612
Wyoming	97	800
Nevada	92	641
Alaska	91	690

The chief outcome of this figure is that the two values are very well correlated in general, and since the regression line has unit slope, the manpower register data are a constant multiple (about antilog $0.5 = 3$) times the number of authors. The scattering about this line indicates states that have more or less than the expected number of authors. For example, Massachusetts, North Carolina, Rhode Island, and Vermont all have more authors and Texas, Delaware, and Idaho fewer authors than would be indicated in proportion to the *National Register* data. It would seem that in the former case there is much basic science in university settings; in the latter case the scientists are concentrated in industry in nonpublishing positions to a greater extent than the national average. In general these deviations would seem to agree with the conventional wisdom about the characteristics of each state, and thereby the method is to some extent validated. It is interesting that the two giant scientific states, New York and

California, are about equal in size and lie right on the regression line without significant deviation.

Since the scientific size of a state must depend on its population we next consider the number of scientific authors per capita as an indicator of scientific "concentration." In figure 9.2 we plot this quantity against one of the several socioeconomic indicators for the states, the per capita personal income. Although income ranges only through a small range of about 0.25 in the logarithm (a factor of ca. 1.7) from lowest to highest, the scatter diagram shows that in general poor states have a small concentration of scientists and rich states a high one. Obvious exceptions to the general trend are Nevada and Alaska where the per capita incomes are artificially boosted by somewhat artificial situations; without these special arrangements Nevada would, it seems, be a state similar to Kentucky or Idaho, and Alaska to Texas, Indiana, Wyoming, or Kansas in quality. Exceptions in the opposite direction are provided by New Mexico and Utah where basic science installations raise the number of scientific authors far above the level expected for states of this wealth. Without Los Alamos, New Mexico would correspond in scientific concentration to Kentucky or Idaho, and thus to Nevada without the casinos. The scatter within the main band of the regression is rather high, but there is nevertheless some correspondence to the placing of a state and the conventional wisdom about its policy in scientific research support and activity.

Lastly, for the states we consider the *relative* rates of growth (i.e., percentage change per annum) of each *relative* to the overall increase in authors for the entire country, i.e., we take the rate of change of the logarithm of number of authors for each state and subtract from that the overall rate of change of the logarithm for

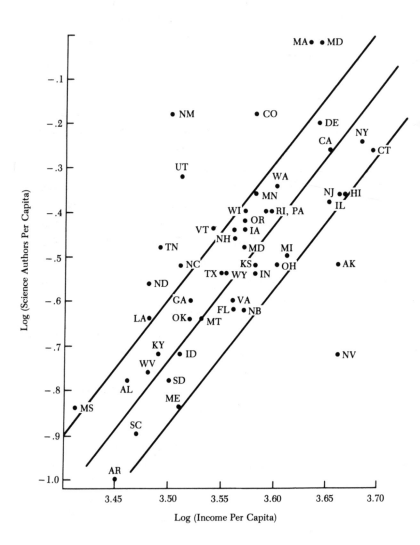

Figure 9.2.

Table 9.2

Logs of Scientific authors per capita (S/P) and personal income per capita (INC/P) for fifty states (District of Columbia and Puerto Rico excluded because of anomalous data)

State	Log (S/P)	Log (INC/P)
Maryland	−0.023081	3.6289
Massachusetts	−0.02652	3.6395
New Mexico	−0.1712	3.4957
Colorado	−0.17975	3.5816
Delaware	−0.2073	3.6359
New York	−0.23051	3.6784
California	−0.25767	3.646
Connecticut	−0.26162	3.6863
Utah	−0.31309	3.5069
Washington	−0.34398	3.6013
Hawaii	−0.36274	3.6558
Minnesota	−0.38684	3.5825
New Jersey	−0.36755	3.6626
Illinois	−0.3746	3.6534
Rhode Island	−0.39794	3.5913
Pennsylvania	−0.40318	3.5941
Wisconsin	−0.40543	3.5674
Oregon	−0.42273	3.5688
Arizona	−0.43622	3.5552
Vermont	−0.44153	3.5397
Iowa	−0.44713	3.5668
New Hampshire	−0.46322	3.5551
Tennessee	−0.47845	3.4893
Missouri	−0.48414	3.5687
Michigan	−0.49592	3.6084
Ohio	−0.51823	3.599
Alaska	−0.52097	3.662
Kansas	−0.52591	3.5824
North Carolina	−0.52821	3.5061
Wyoming	−0.53437	3.551
Texas	−0.54314	3.5479
Indiana	−0.54524	3.5776
North Dakota	−0.56827	3.4764
Georgia	−0.59141	3.5227
Virginia	−0.59172	3.5571
Florida	−0.62525	3.5613
Nebraska	−0.62985	3.5741

Table 9.2 *Continued*

State	Log (S/P)	Log (INC/P)
Montana	–0.63184	3.5288
Louisiana	–0.63718	3.4842
Oklahoma	–0.64689	3.5201
Kentucky	–0.71045	3.4876
Nevada	–0.72552	3.6592
Idaho	–0.72598	3.5105
West Virginia	–0.7601	3.4802
South Dakota	–0.7704	3.5004
Alabama	–0.77139	3.4553
Mississippi	–0.83656	3.4108
Maine	–0.83902	3.5128
South Carolina	–0.90025	3.4678
Arkansas	–0.99172	3.4458

the whole United States. A scatter plot of this quantity against the sizes of the individual states is shown in figure 9.3, the logarithms of the sizes being shown relative to the average logarithmic value for all states. Though the scatter is quite large, it is evident that in general the large states grow below the average rate, and the small states grow most rapidly. For the largest states, indeed, one must suspect that New York, California, and Massachusetts, with relative growth rates about 7 units below average, are stationary in absolute terms. That is to say, the lower limit of the scatter probably corresponds to a zero growth rate, and the middle of the scatter to the average rate for the nation as a whole. Looking again at the most exceptional cases, we find that for states of their sizes we have spectacularly rapid growth rates in the District of Columbia, Texas, and Georgia, and at the other end of the scale, surprisingly little growth for Delaware and Puerto Rico. It is

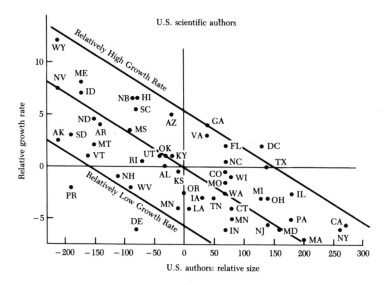

Figure 9.3. U.S. Scientific Authors

surprising that the general differential effect is so large between the large, concentrated states that are now almost stationary and the small, still scientifically diluted states that continue to grow. The effect is such that at a rough estimate it would be less than twenty years at the present rates for the United States to become uniformly concentrated in per capita scientists except for the local anomalies already pointed out.

Table 9.3
Logs of relative size and relative growth rate for states, ranked by growth rates

State	Relative Size (Log)	Relative Growth Rate
Wyoming	−206.41	12.14
Maine	−166.53	7.9408
Nevada	−210.76	7.4583
Idaho	−173.53	6.7827
Hawaii	−82.253	6.7251
Nebraska	−78.214	6.4071
South Carolina	−84.821	5.4018
Arizona	−15.852	5.2005
North Dakota	−151.78	4.3011
Arkansas	−135.4	4.2235
Georgia	43.553	4.0996
Mississippi	−85.692	3.4993
South Dakota	−190.69	2.9675
Virginia	44.787	2.8967
Alaska	−212.81	2.4175
Florida	74.929	2.0747
Montana	−154.64	1.9787
District of Columbia	132.55	1.8684
Utah	−38.936	1.1672
Kentucky	−19.363	0.95845
Oklahoma	−27.657	0.88161
Vermont	−155.32	0.78785
Rhode Island	−69.296	0.45742
North Carolina	68.292	0.37729
Alabama	−26.53	0.13418
Texas	143.87	−0.2222
Kansas	−12.681	−0.31965
Colorado	65.134	−0.62909
Wisconsin	82.58	−1.1359
New Hampshire	−109.56	−1.1866
Missouri	70.156	−1.5221
West Virginia	−92.195	−2.0024
Puerto Rico	−186.47	−2.0283
Oregon	3.8492	−2.5202
Illinois	181.92	−2.6811
Washington	70.808	−2.7214
Ohio	144.62	−2.8697

Table 9.3 *Continued*

State	relative Size (Log)	Relative Growth Rate
Tennessee	53.931	-3.1296
Michigan	131.51	-3.1507
Iowa	28.3	-3.1988
Louisiana	9.9579	-3.9305
New Mexico	-10.484	3.9927
Connecticut	78.021	-4.0001
Pennsylvania	181.28	-4.8224
Minnesota	76.501	-5.1823
California	267.37	-5.3157
New Jersey	139.7	-5.5497
New York	264.38	-5.8153
Delaware	-80.495	-5.8733
Maryland	158.72	-6.1387
Indiana	66.539	-6.2227
Massachusetts	195.11	-6.9844

NATIONS OF THE WORLD

The author counts for the nations of the world are now presented for the first time in forms that show very much more regularity and less scatter than has, we think, ever been demonstrated. Part of this improvement is due to the smaller random noise of a seven-year geometric mean rather than the single year value used previously.[1]

Another big step forward has been the choice of a more appropriate indicator for size and development of the various nations. In the past the measure used has usually been the gross national product (or per capita GNP) of the country, but this is notoriously unreliable and of questionable validity, particularly for the planned economies of the world where the conversion factors must be arbitrary. Partly because of its universality and partly because it has been found to correlate very well and in a similar way to science indicators, we now propose an energy measure, the number of kilowatt-hours of electricity used in the nation as recorded in the standard *Statistical Abstracts* (we used the 1973 volume which gives data for 1971).

In figure 9.4 we show such a scatter plot for the absolute logarithmic sizes of scientific authors against kilowatt-hours. The main band of regression is very clear with only Israel as a highly anomalous nation, followed by Lebanon, Ceylon, Ethiopia, and Hungary with somewhat less deviation above the main trend. In all such cases we are dealing with countries having many more scientific authors than would be expected from the magnitude of energy consumption. Stating it in an alternative fashion, the same countries have much less energy consumption (i.e., technological development) than would be expected from their scientific contribution. In general there is a tendency for former British Commonwealth countries to be high in devotion to "pure" science, and for Spanish-speaking countries to be low in science, perhaps because of the inadequacy of international scientific literature in what is nevertheless a major world language.

Table 9.4

Logarithmic values of numbers of scientific authors (geometric mean 1967–1973) and electrical consumption of major nations for which good data are available, ranked by authors

Country	Log S	Log KWHR
U.S.	4.9355	6.2349
U.K.	4.2891	5.4084
USSR	4.198	5.9033
Germany, Fed. Rep.	4.056	5.4045
France	4.0007	5.1732
Japan	3.9096	5.5862
Canada	3.8847	5.3354
India	3.6589	4.7809
Australia	3.5687	4.7566
Italy	3.5462	5.0964
Switzerland	3.3948	4.5157
Sweden	3.378	4.8231

Table 9.4 *Continued*

Country	Log S	Log KWHR
Czechoslovakia	3.3547	4.6743
Netherlands	3.3404	4.6523
Poland	3.2572	4.8444
Germany, Dem. Rep.	3.2541	4.8415
Israel	3.2315	3.883
Hungary	3.1578	4.1758
Belgium	3.1532	4.4996
Denmark	3.0402	4.244
Austria	2.989	4.4586
Norway	2.8943	4.7989
Romania	2.8871	4.5961
Finland	2.8603	4.3189
South Africa	2.7993	4.722
New Zealand	2.7716	4.1817
Argentina	2.7007	4.3733
Bulgaria	2.6981	4.3226
Spain	2.6803	4.7918
Yugoslavia	2.6522	4.47
U.A.R.	2.617	3.8803
Brazil	2.5775	4.7075
Ireland	2.4843	3.8017
Mexico	2.3892	4.4958
Chile	2.3729	3.9306
Greece	2.3304	3.9257
Venezuela	2.2967	4.1332
Taiwan	2.1303	4.1321
Pakistan	2.1271	3.8493
Turkey	2.0086	3.9904
Lebanon	1.9243	3.1383
Thailand	1.8513	3.6575
Portugal	1.8325	3.8929
Philippines	1.7404	3.9378
Ghana	1.699	3.4689
Colombia	1.6902	3.942
Iraq	1.6435	3.2853
Peru	1.6335	3.7262

Table 9.4 *Continued*

Country	Log S	Log KWHR
Ceylon	1.6128	2.9542
Korea	1.5441	4.0395
Uruguay	1.5315	3.2819
Costa Rica	1.301	3.0599
Ethiopia	1.301	2.7672
Guatemala	1.2041	2.8921
Iceland	1.1461	3.2047
Tunisia	1.1139	2.9479
Cuba	1.0792	3.6098
Bolivia	0.69897	2.9191
Burma	0.69897	2.7987
Ecuador	0.69897	3.1593
Luxembourg	0.69897	3.3672
El Salvador	0.60206	2.871
Liberia	0.47712	2.8129
Panama	0.47712	2.9805
Syria	0.47712	3.0208
Honduras	0.30103	2.4914
Nicaragua	0.30103	2.6767
Paraguay	0.30103	2.3729

The same basic data is shown on a per capita basis in figure 9.5, and once again the correlation is excellent. The state of Israel still shows a much higher concentration of science relative to energy use, but its case is balanced now by Norway where there is a superabundance of hydroelectric energy. In general the developed countries of Western Europe lie within a small circle of scattering at the top end of the regression strip, and the Eastern European nations together with Japan form a similar circle just below this. Again, the former British Empire countries are high on the strip and the Spanish-speaking lands (except for Argentina) rather low. We are at present testing the hypothesis that if one goes from such a synchronous study to its diachronous equivalent,

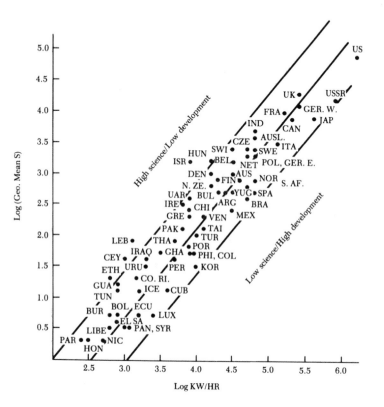

Figure 9.4. Absolute Sizes

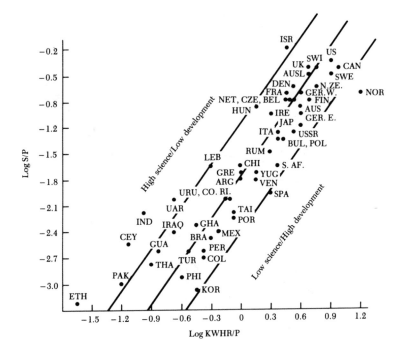

Figure 9.5. Per Capita Sizes

tracing the paths of countries as a function of time, it will be seen that nations move up, along the regression strip, fast at the lower end, more slowly as they crowd up to and ultimately reach and remain static at the top level of most-developed countries.

Some evidence for this is given by figure 9.6, which is a scatter plot for nations matching that of figure 9.3 for the states of the United States. Once more, with the exception of the smallest countries (and for some reason, also Uruguay) whose growth rates

are probably much affected by random noise phenomena, there
is a clear tendency for relative growth rates to decrease with

Table 9.5
Logs of scientific authors per capita and kilowatt-hours of electricity per
capita for the major nations for which good data are available

Country	Log(S/P)	Log(KWH/P)
Israel	–0.23242	0.41914
U.S.	–0.37601	0.9234
Switzerland	–0.40247	0.71841
Canada	–0.44405	1.0066
U.K.	–0.45704	0.66232
Sweden	–0.52722	0.91789
Australia	–0.52859	0.65937
Denmark	–0.65264	0.55118
New Zealand	–0.67712	0.73297
Norway	–0.69452	1.21
France	–0.70487	0.46757
Germany, Fed. Rep.	–0.72688	0.62165
Netherlands	–0.7745	0.53734
Czechoslovakia	–0.8008	0.51879
Finland	–0.81083	0.64777
Belgium	–0.83177	0.51467
Hungary	–0.85156	0.16648
Austria	–0.87964	0.58996
Germany, Dem. Rep.	–0.98272	0.6047
Ireland	–0.98552	0.33186
Japan	–1.1049	0.57167
USSR	–1.174	0.53126
Italy	–1.1827	0.3675
Bulgaria	–1.2308	0.39364
Poland	–1.2472	0.34001
Romania	–1.4194	0.2896
Lebanon	–1.5213	–0.3073
South Africa	–1.5321	0.39052
Chile	–1.5745	–0.01679
Greece	–1.6136	–0.018316
Yugoslavia	–1.643	0.17472
Argentina	–1.665	0.0076599

Table 9.5 *Continued*

Country	Log (S/P)	Log(KWH/P)
Venezuela	-1.7204	0.11615
Spain	-1.8483	0.26315
U.A.R.	-1.9058	-0.64254
Uruguay	-1.9294	-0.17896
Costa Rica	-1.9395	-0.18061
Taiwan	-2.0394	-0.037636
India	-2.0734	-0.95137
Portugal	-2.1313	-0.070859
Ghana	-2.2375	-0.46758
Mexico	-2.3018	-0.19514
Iraq	-2.3315	-0.68964
Brazil	-2.3928	-0.26283
Ceylon	-2.4845	-1.143
Peru	-2.4998	-0.40698
Guatemala	-2.511	-0.82307
Turkey	-2.5383	-0.55653
Colombia	-2.6345	-0.38269
Thailand	-2.685	-0.87877
Philippines	-2.8261	-0.62862
Pakistan	-2.9305	-1.2083
Korea	-2.9518	-0.45633
Ethiopia	-3.0904	-1.6243

increasing size of the nation. Probably, as before, the lowest general level of growth rate, ca. -11, corresponds to a zero growth rate on the absolute scale, and the zero of the relative growth rate corresponds to the absolute world average of ca. 7 percent increase per annum. Of the developed nations, only Canada preserves a high relative growth comparable with that of the much smaller developing countries. Yet again, it is unexpected that the trend toward uniform concentration is so rapid that, if present rates were continued, the entropic change would be complete within about twenty years. At any rate, this provides, so far as we can see, the first positive evidence that the famous

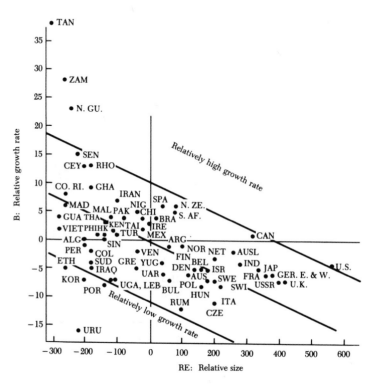

Figure 9.6. World Scientific Authors

developmental gap between rich countries and poor is certainly
not increasing but rather decreasing quite rapidly.

Table 9.6

Log of relative number of scientific authors and relative growth rate for
major nations for which good data are available—ranked by relative
growth rate

Nation	Log Relative Size	Relative Growth Rate
Tanzania	−298.74	38.761
Zambia	−261.62	30.709
New Guinea	−252.57	24.69
Senegal	−235.18	17.713
Ceylon	−198.08	13.624
Rhodesia	−181.11	13.14
Costa Rica	−269.94	9.5376
Ghana	−177.97	8.7949
Iran	−98.871	6.7931
Madagascar	−267.88	6.7848
Spain	48.161	5.9927
New Zealand	69.149	5.9062
Guatemala	−294.02	5.2397
South Africa	75.578	4.7332
Nigeria	−47.36	4.6573
Malaysia	−121.34	4.4342
Brazil	24.373	4.074
Chile	−22.689	4.0336
Pakistan	−79.059	3.6082
Vietnam	−289.54	3.4278
Thailand	−142.28	3.1492
Ireland	2.8207	3.0774
Kenya	−130.01	2.7296
Taiwan	−78.295	2.3211
Hong Kong	−148.33	1.4744
Mexico	−18.866	1.35
Turkey	−106.27	1.0676
Canada	325.42	0.9816
Philippines	−168.31	0.60189
Singapore	−136.91	0.31299
Algeria	−197.45	−0.10735
Argentina	52.692	−0.87533

Table 9.6 *Continued*

Nation	Log Relative Size	Relative Growth Rate
Peru	-193.65	-1.0291
Norway	97.341	-1.2413
Venezuela	-40.285	-1.8167
Colombia	-179.77	-2.0965
Australia	252.65	-2.497
Netherlands	200.11	-3.5659
Finland	89.563	-4.217
U.S.	567.37	-4.3999
Sudan	-178.25	-4.4387
Yugoslavia	41.713	-4.5733
India	273.41	-4.6603
Israel	175.02	-4.9153
Iraq	-189.92	-4.9387
Ethiopia	-267.95	-4.9502
Belgium	157	-5.2723
Greece	-32.627	-5.5256
Japan	331.14	-5.5744
Denmark	130.98	-5.6776
U.A.R.	33.575	-5.7937
Germany, Dem. Rep.	180.18	-6.1903
Austria	119.2	-6.3258
Germany, Fed. Rep.	364.84	-6.4151
France	352.13	-6.5336
USSR	397.56	-6.862
Lebanon	-126.45	-6.905
Uganda	-131.23	-6.948
Korea Rep.	-212.98	-6.9702
U.K.	418.52	-6.9947
Poland	180.91	-7.5559
Bulgaria	52.238	-7.5688
Sweden	208.76	-7.7098
Switzerland	212.6	-7.7973
Hungary	158.02	-7.8229
Portugal	-147.74	-7.8608
Italy	247.47	-9.3252
Czechoslovakia	203.39	-11.054

Table 9.6 *Continued*

State	1967	1968	1969	1970	1971	1972	1973
Romania			95.689			−12.298	
Uruguay			−216.02			−16.415	

Table 9.7
Number of *WIPIS* authors in each state of United States by year
U.S. Scientific Authors

State	1967	1968	1969	1970	1971	1972	1973
Alabama	309	430	564	645	656	822	882
Alaska	55	56	60	118	118	147	132
Arizona	323	343	605	867	773	996	1085
Arkansas	106	122	158	223	243	294	346
California	7270	8444	10929	12819	11612	14033	14118
Colorado	829	984	1349	1832	1679	2028	2049
Connecticut	1031	1237	1660	1989	1727	2164	2203
Delaware	222	301	318	363	353	439	440
Dist. of Columbia	1506	1725	2701	3717	3454	4201	4166
Florida	884	947	1540	1940	1895	2386	2469
Georgia	559	723	1104	1453	1445	1740	1905
Hawaii	136	206	314	448	387	523	583
Idaho	68	73	104	161	168	265	211
Illinois	2795	3142	4926	5772	5276	6027	6291
Indiana	885	1042	1618	1979	1803	2200	1325
Iowa	591	751	999	1206	1168	1290	1325
Kansas	375	425	692	826	756	898	980
Kentucky	346	391	627	716	762	851	964
Louisiana	546	591	848	972	903	1103	1116
Maine	70	72	119	198	177	212	286
Maryland	2545	3060	3497	4177	3838	4779	4720
Massachusetts	3741	4340	5312	6143	5347	6631	6695
Michigan	1714	1970	2967	3366	3126	3677	3781
Minnesota	1086	1259	1585	1927	1680	2052	2165
Mississippi	168	200	262	416	401	491	507
Missouri	872	1038	1538	1862	1806	2059	2074
Montana	80	103	161	200	208	215	247

Table 9.7 *Continued*

State	1967	1968	1969	1970	1971	1972	1973
Nebraska	165	181	322	454	441	535	599
Nevada	39	53	81	126	113	154	157
New Hampshire	154	181	222	297	284	358	368
New Jersey	1964	2438	3059	3569	3397	3849	3805
New Mexico	442	512	674	765	789	752	1021
New York	7139	7991	11286	12401	11097	13445	13467
North Carolina	870	1024	1343	1739	1682	2160	2321
North Dakota	84	102	129	230	212	246	270
Ohio	1958	2293	3187	3843	3633	4201	4373
Oklahoma	317	372	513	766	630	881	826
Oregon	494	506	822	962	839	1105	1052
Pennsylvania	3108	3429	4551	5532	4855	6042	6070
Puerto Rico	64	84	115	167	132	143	162
Rhode Island	203	256	390	445	411	533	583
South Carolina	168	197	274	361	377	540	583
South Dakota	51	66	104	195	138	163	153
Tennessee	818	971	1265	1421	1389	1822	1777
Texas	1922	2151	2842	3741	3587	4626	4774
Utah	287	319	493	616	589	769	766
Vermont	91	94	166	200	180	224	244
Virginia	616	761	1110	1328	1439	1764	1930
Washington	927	1097	1533	1878	1660	2029	2122
West Virginia	180	189	316	398	344	376	419
Wisconsin	973	1153	1762	2248	1892	2312	2454
Wyoming	38	42	80	147	143	160	182

Table 9.8
Numbers of authors in major nations by year. *The numbers for
"Germany, Dem. Rep." and "Germany, Fed. Rep." 1967 to 1971 have
been calcualted from the proportions of their numbers in 1972 and
1973

World Scientific Authors

	1967	1968	1969	1970	1971	1972	1973
Algeria	22	32	31	47	52	60	62
Argentina	299	348	491	488	505	760	834
Australia	2038	2728	3709	4290	3914	5236	5280

Table 9.8 Continued

	1967	1968	1969	1970	1971	1972	1973
Austria	646	851	882	966	1107	1178	1374
Belgium	924	1171	1320	1459	1473	1902	2027
Brazil	207	214	327	439	388	607	734
Bulgaria	376	433	409	529	486	672	672
Canada	3997	5085	6776	9056	9206	11586	11731
Ceylon	11	27	35	51	52	81	85
Chile	113	134	212	273	340	377	363
Colombia	30	38	34	58	60	78	65
Costa Rica	10	7	14	38	28	35	34
Czechoslovakia	1718	2052	2166	2598	2161	2713	2616
Denmark	728	938	952	1182	1091	1449	1576
Ethiopia	11	21	18	26	16	25	33
Finland	447	609	624	810	731	975	1075
France	6862	8234	9140	10437	10585	13515	13125
Germany, Dem. Rep.*	1144	1517	1719	1962	1860	2183	2523
Germany, Fed. Rep.*	7254	9616	10899	12439	11791	14646	15085
Ghana	24	31	35	52	54	87	122
Greece	147	166	214	202	220	273	321
Guatemala	7	8	9	38	26	22	21
Hong Kong	30	47	70	74	90	87	108
Hungary	1039	1171	1289	1674	1461	1780	1832
India	2882	3512	4078	5116	5144	6189	6086
Iran	52	57	95	127	145	201	189
Iraq	32	29	43	45	52	60	60
Ireland	156	208	223	385	356	445	551
Israel	1125	1297	1596	1815	1787	2304	2401
Italy	2733	3058	3320	3682	3279	4064	4890
Japan	5202	6825	7659	8433	8473	10158	11794
Kenya	34	60	59	126	100	130	113
Korea Rep.	22	36	28	41	37	39	51
Lebanon	58	71	68	95	93	109	106
Madagascar	9	9	20	27	32	32	32
Malaysia	41	54	81	100	105	140	155
Mexico	152	180	183	231	246	391	479
Netherlands	1412	1623	1958	2273	2451	3050	3172
New Guinea	9	7	12	30	43	58	74

Table 9.8 *Continued*

	1967	1968	1969	1970	1971	1972	1973
New Zealand	253	319	537	828	755	972	959
Nigeria	97	111	140	203	242	288	340
Norway	432	593	703	864	856	1135	1202
Pakistan	68	83	108	145	193	218	212
Peru	32	27	37	50	32	46	110
Philippines	32	35	47	66	59	80	93
Poland	1305	1534	1539	1941	1970	2294	2335
Portugal	51	66	51	74	55	97	95
Rhodesia	25	18	36	58	51	114	114
Romania	557	724	771	911	851	849	791
Senegal	4	9	106	30	53	63	37
Singapore	38	53	64	95	92	114	107
Spain	277	272	370	477	517	737	1147
Sudan	26	33	59	72	58	61	59
South Africa	340	425	450	629	691	1163	1205
Sweden	1652	2060	2203	2530	2524	3027	3059
Switzerland	1768	2091	2389	2625	2402	3068	3395
Taiwan	72	80	122	164	155	206	226
Tanzania	10	1	4	27	45	63	54
Thailand	41	46	67	61	78	94	167
Turkey	59	63	86	125	112	157	167
Uganda	57	67	67	83	89	104	104
U.A.R.	293	304	380	370	442	547	559
U.K.	13103	15659	18657	21955	19976	25114	25026
U.S.	52184	60442	83203	114676	92014	111717	114279
USSR	10525	13928	13368	17457	17215	20158	20494
Uruguay	34	33	38	28	34	36	37
Venezuela	82	326	105	347	137	193	462
Vietnam	8	8	12	27	33	23	20
Yugoslavia	288	364	396	483	481	556	633
Zambia	6	4	15	38	40	70	58

10

Studies in Scientometrics, Part 1: Transience and Continuance in Scientific Authorship

Many of the richest areas for research in the sociology of science depend upon some understanding of what may be called the actuarial statistics of the scientific community. One needs to know the dynamical processes which govern emergence, survival, and disappearance within that community. These determine the structure of the group by age, status, productivity, reputation, and professional ties. Such studies have many of the same strengths and limitations as actuarial methods in demography and life insurance. Useful calculations may be made about the population at large, but the bearing of the life of any individual remains statistical rather than causal. The purpose of this investigation is to uncover the facts and regularities which will require some theoretical explanation. Undoubtedly the most important phenomenon, hitherto not well recognized, is that at any given time a large number of those working at the research front are transients whose names have never appeared before and will not appear again in the record. The point has obvious application to the natural history of scientific careers, and it is also of fundamental importance to the analysis of manpower data in the sciences, since only part of the research labor force can be considered as stable.

Previous work in this area has usually been based upon hand

Coauthored by Suha Gürsey.

or machine counts that have been limited to a single nation, a scientific specialty, or just one journal or scientific institution. The results have always been of questionable generality because of the possible strong idiosyncracies of these special groups and also because of the large general movement that exists across the boundaries of such groups as people change jobs and migrate through fields. We have been fortunate in having at our disposal data emerging as a by-product from the machine handling of a uniquely comprehensive and worldwide coverage of the literature in all fields of basic and applied science. For this reason the results are relatively free of local idiosyncracies and are of general applicability to the scientific community.

The data bank for this study was based upon volumes published by the Institute for Scientific Information, including several years of output of the *Science Citation Index*, with its indexes of source authors and cited authors, and the annual volumes of *Who Is Publishing in Science*, which is derived from the weekly editions of *Current Contents*.[1] These indexes cover all the principal journals. The criterion for inclusion is that of usage by the scientific community; many known journals are excluded, particularly local and domestic periodicals, but only because scientists do not cite them at all in subsequent research publications.

To manipulate the entire data bank, derived from more than 2,000 journals and amounting to millions of citations each year, would have been far too costly and complicated and would have excluded any possibility of hand collation and editing, which is essential in work of this kind. We therefore devised the simple technique of generating a small but random intercomparable sample of all indexes by taking only those authors whose names fell into a limited slice of the alphabetic listing. The range was chosen after several trials to begin with a person whose work was known to us and to extend for about 100 names in the first index covered.

We were careful that the selection did not happen to include, so far as could be detected, any wildly pathological bias toward authors of any particular national or linguistic group. It contained no well-known family active in science, nor any common surname that would pick up several different individuals for a single set of initials. Our final selection was the slice from Pais, A. to Palecek, M., which corresponds to about 0.22 ± 0.02 percent of the entire author index in any of the many lists searched. By studying in detail a group of about 500 individuals drawn in this way from indexes dating from the period 1964–1970 we were therefore sampling a total population of a little more than a million scientific authors in all the countries of the world. This is a size consistent with most known estimates of the total world population of all research scientists and engineers.

The names of every author on each paper are fed into the source indexes, so each annual record contains a listing of the scientific population, new and old, whose names are on the bylines of all papers published during the previous year. The total number of different names so recorded in the seven years was 506, and the annual totals increased in the usual exponential fashion, starting from 96 in 1964 and almost doubling to 187 in 1970. It should be noted that this doubling in seven years (a growth rate of 10 percent per annum) is only partly due to actual increase in the scientific population; probably some 3 or 4 percent of the growth is due to increased coverge of the I.S.I. services as journals were added to the source list.

The raw data emerging from this study are displayed in table 10.1, which shows the number of authors who were listed in each of the possible combinations of years.

It can be seen immediately that there is a certain tendency toward extreme behavior; authors seem to lean toward either publishing in only one single year, or publishing in all available

Table 10.1

Years	Number of Authors	Years	Number of Authors	Years	Number of Authors	Years	Number of Authors
64	25	64/65/66	1	64/65/66/67	2	64/65/66/67/68	3
65	32	64/65/67	4	64/65/66/68		64/65/66/67/69	1
66	33	64/65/68		64/65/66/69	1	64/65/66/67/70	
67	41	64/65/69	2	64/65/66/70		64/65/66/68/69	1
68	42	64/65/70	1	64/65/67/68		64/65/66/68/70	1
69	49	64/66/67		64/65/67/69		64/65/66/69/70	
70	59	64/66/68	1	64/65/67/70		64/65/67/68/69	1
		64/66/69		64/65/68/69		64/65/67/68/70	1
64/65	3	64/66/70	1	64/65/68/70		64/65/67/69/70	2
64/66	3	64/67/68		64/65/69/70		64/65/68/69/70	
64/67	2	64/67/69		64/66/67/68	2	64/66/67/68/69	1
64/68		64/67/70	1	64/66/67/69		64/66/67/68/70	
64/69		64/68/69		64/66/67/70		64/66/67/69/70	2
64/70	1	64/68/70		64/66/68/69	1	64/66/68/69/70	
65/66	4	64/69/70	1	64/66/68/70	1	64/67/68/69/70	
65/67	2	65/66/67	3	64/66/69/70		65/66/67/68/69	1
65/68	2	65/66/68	1	64/67/68/69	1	65/66/67/68/70	2
65/69	2	65/66/69		64/67/68/70		65/66/67/69/70	
65/70	1	65/66/70	2	64/67/69/70	1	65/66/68/69/70	4
66/67	4	65/67/68		64/68/69/70	2	65/67/68/69/70	2
66/68	4	65/67/69		65/66/67/68	2	66/67/68/69/70	8
66/69	2	65/67/70	2	65/66/67/69	1		
66/70	9	65/68/69	1	65/66/67/70		64/65/66/67/68/69	
67/68	7	65/68/70		65/66/68/69		64/65/66/67/68/70	1
67/69	7	65/69/70		65/66/68/70		64/65/66/67/69/70	1
67/70	5	66/67/68	3	65/66/69/70	3	64/65/66/68/69/70	1
68/69	8	66/67/69	2	65/67/68/69	1	64/65/67/68/69/70	2
68/70	4	66/67/70	2	65/67/68/70	1	64/66/67/68/69/70	2
69/70	16	66/68/69	1	65/67/69/70	1	65/66/67/68/69/70	2
		66/68/70	1	65/68/69/70			
		66/69/70	4	66/67/68/69	3	64/65/66/67/68/69/70	19
		67/68/69	6	66/67/68/70			
		67/68/70	1	66/67/69/70	1		
		67/69/70	2	66/68/69/70	2		
		68/69/70	6	67/68/69/70	6		

years. The results can also be summarized in a diagram which shows for each year the number of authors publishing in that year, the contribution to this by authors publishing before and not before, and the number of authors who publish subsequently and not subsequently. Figure 10.1 shows this flow, the authors being broken down into those who skip a year or more before or after the year of publication in question. The figure also shows separately those who fall in the category of not having been recorded before or subsequently.

In each year studied those in this latter category of authors who

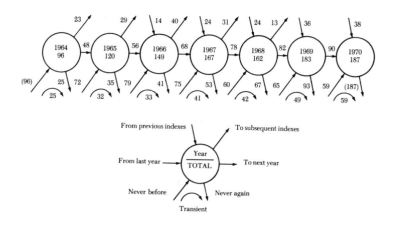

Figure 10.1. Flow of source authors through seven consecutive
annual indexes.

have never been heard of before and never are heard of again amount to about 25 percent of those recorded for that year. They are surprisingly numerous, and, of course, the proportion of them in the total population increases with the span of years considered because the rest of the authors' names occur in two or more years. In the whole period of seven years there are 281 names that occur in a single source index out of the 506 authors listed, a fraction of 56 percent of the poplulation. We shall call this phenomenon transience and such authors transients. It is important to note that these cannot be people who have migrated to a different field of research, for the corpus includes all publishable fields of science and technology, all institutions and countries.

The opposite behavior pattern to transience, that of authors whose names tend to appear year after year in every index of the record, we shall call continuance. For the seven-year period there are 19 such authors, and though they are but a small proportion of the total population of authors for all those years together, they constitute 20 percent of the pool of authors recorded at the beginning of the period in 1964. Such strongly continuing authors are clearly those who normally publish much more than a single paper a year, so that their chance of skipping a year is very small. There must be many more whose continuance is slightly less because of somewhat smaller production rate or an occasional sabbatical or period of work on a book or extensive monograph.

By the same token we shall need to weaken the definition of transience, for there must be some otherwise transient publishers whose single research front production happens to appear in two or even three papers that fate and publishing time lags decree shall come out in journals falling into two adjacent annual gatherings. The basic fact is, however, that nearly half the authors on an annual index are either strongly transient (25 percent) or strongly continuant (20 percent), so that they will continue to publish each year for the next six years. The remaining 55 percent of the annual author list consists partly of those who are more weakly transient or continuant, and partly of those who are newly recruited or newly terminated from the continuing community.

To derive a better quantitative analysis of the author flow pattern it is strategic to base an initial approximation upon the central year of the series, 1967, for which one has records that can detect authors whose continuance is so weak that they reappear after as much as two consecutive years of skipping publication. The general pattern of flow is shown in figure 10.2.

The transients account for 25 percent of the population, or to put it into demographic terms, there is a 25 percent infant mortality, over and above a birthrate of 20 percent and a death rate of 10 percent. To put it another way, there is a total birthrate of 45 percent and a death rate of 35 percent, which overlap to give the transients. Just over half the authors in this year have been previously recorded, 41 percent in the immediately prior year, another 11 percent after skipping one year, and 3 percent after a skip of two years, the total being 55 percent, but of these 10

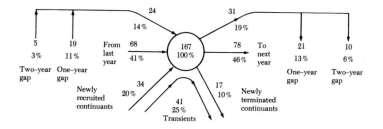

Figure 10.2. Empirical data for flow of source authors in 1967.

percent terminate their publishing in this year, so that only 45% of the previously established authors will continue. These are augmented by 20 percent of the population who are new recruits, and the resultant 65 percent go on to publish, 46 percent of them in the next year, 13 percent after a gap of one year, and another 6 percent after a two-year gap.

To emphasize the magnitude of the transience phenomenon, it may be noted that in this first approximation they constitute 25 percent out of the total birthrate of 45 percent, a fraction of 0.56 of all births; similarly they are 0.71 of all the death rate. These figures must, however, be modified a little because there exist small groups of authors who publish again after skipping more than two years of indexes. The effect of this is that a small amount of those who were here considered in the total birthrate must in fact be reckoned as authors reappearing after skipping three or more years, and correspondingly part of the assumed total death rate is due to authors who will reappear after a gap of three or more years. Fortunately, skipping is a relatively uncommon phenomenon.

The data for the group of authors publishing in 1967 show that 66 percent had a continuous record of publication, a further 22 percent had a single gap, and 10 percent had two gaps, leaving only 2 percent with a gap of three or more years in the entire interval of seven years.

Since, in fact, a full seven-year period was studied, it is possible to get data on larger gaps directly by starting from the first and the last years in the series. The results are shown in table 10.2.

Clearly the frequency of long gap records is so low that it is heavily influenced by random noise. Taking the average of the 1964 and 1970 data, we shall assume that gaps of three years occur for 4 percent of the authors and gaps of four and five or more years with 1 percent frequency for each. The series decreases so

Table 10.2

a) Of the 96 authors publishing in 1964		%
25 do not publish again		26
48 proceed with no skips to	1965	50
14 skip 1 year, publish in	1966	15
5 " 2 years "	1967	5
2 " 3 years "	1968	2
1 " 4 years "	1969	1
1 " 5 years "	1970	1

b) Of the 187 authors publishing in 1970		%
59 have not published before		32
90 published with no skips in	1969	48
13 skipped 1 year, published in	1968	7
10 " 2 years "	1967	5
12 " 3 years "	1966	6
2 " 4 years "	1965	1
1 " 5 years "	1964	1

c) Summary of data %	1964 and after	1970 and before	1967 and before	1967 and after
Not publishing again	26	32	45	35
Publish next year	50	48	41	46
1 skip	15	7	11	13
2	5	5	3	6
3	2	6	—	—
4	1	1	—	—
5	1	1	—	—

rapidly that we may safely suppose gaps of more than five years to be vanishingly rare; an author who has not published for the last six years may be considered as terminated. The 6 percent of authors having long gap records must now be subtracted from the previous estimates of total birthrate and death rate, so that we must now suppose the annual recruitment to be $45 - 6 = 39\%$, and the annual termination to be $35 - 6 = 29\%$. The effect of this correction upon the magnitude of the overlapping area of transience is not immediately clear. The newly identified 6 percent of large gap authors may be considered as those who were apparently new recruits and terminators among the continuants, or as being equally divided among both classes of either the input or the output. The results can be seen in table 10.3.

For our second approximation we shall take the mean of these four possibilities, which is, in fact, given by the third line, in which transients amount to 22 percent of any annual index of authors. They outnumber perceptibly the new recruits to the community of continuing authors, and they are more than double the natural rate of increase (birthrate minus death rate) for the scientific community. To put it very roughly, for every increase of 1 author in the continuant population it is necessary that about 4 new authors come into being; of these 1 replaces somebody who ceases to publish, 2 represent the infant mortality of those who arrive and depart from the research front in the same year, and the fourth survives for a greater or lesser lifetime of publication. Though this crude model needs much refinement and correction, the fact is clear that recruitment to the relatively stable community of continuing scientific authors proceeds in two stages. Reaching the research front and producing one's first paper—a process institutionalized in the United States by the Ph.D.—is only a first step. The large majority of those attaining this step are destined to proceed elsewhere than further research front publication in any scientific or technological field. Only a fraction, perhaps a

Table 10.3

Large gap authors	Recruits	Terminators	Transients
	(Percentages)		
all transients	20	10	19
nontransients	14	4	25
distributed by all births	17	7	22
distributed by all deaths	18	8	21

quarter, of those reaching the research front cross the barrier from transience to continuance in production of scientific papers.

One obvious fact in the above model is that the 10 percent per annum growth rate in numbers of authors reflects the growth of the ISI indexes over and above that of the population of authors. The average world rate of production of authors should almost certainly be nearer to 7 percent per annum, resulting in a doubling period of about 10 years. A more serious fault is that we have only rough empirical values for the incidence of overlapping and gaps in the sort of year-to-year presented in table 10.2.

To develop a better model, let us first consider the case in which there exists a fixed and stable continuant population C whose members year after year have a fixed probability p of publishing during that year, and a corresponding probability $1 - p$ of not publishing. The number publishing in two consecutive years will be Cp^2, and the number publishing after a gap of one or more years will be $Cp(1 - p)$. Fitting this model to our empirical data as presented in figure 10.2 and modified by the transfer of 6 percent from newcomers to those reappearing after long gaps, we find that the number coming from the previous year is 41 percent, and the number reappearing after short and long gaps is 20 percent. Hence $Cp^2 = 41\%$ and $Cp(1 - p) = 20\%$, from which $p = 0.67$ and $C = 91\%$. We may take the convenient approximate values of $p = \frac{2}{3}$ and $1 - p = \frac{1}{3}$ and compute from this that the numbers appearing after gaps of one, two, three, and four years should be 14 percent, 5 percent, 2 percent, 0.5 percent, all in rather good agreement with the empirical data of (c) in table 10.2.

Our model therefore requires that in addition to the 39 newcomers and the 61 continuant publishers who exist among every 100 authors publishing in a given year, there are another 30 continuants who did not happen to publish in the year in question.

The total "scientific" population of possible publishers is therefore 130, and 1.3 times the number that actually publish in that year, but 39 of these are newcoming transients and recruits, and of those remaining only ⅔ actually publish during the year.

Considering next the fact that our model must not be static, but subject to exponential growth, we note that by the end of the year in question the 91 publishing and nonpublishing continuants have been augmented by the natural increase of 10 which is the excess of newcomers over terminators. There will therefore be now 101 continuants of whom 67 publish during the next year and 33 do not. Of the 67 there will be 45 continuing directly to the subsequent year and 22 who reappear after publishing gaps of one or more years. Again, this is in good agreement with the right-hand side of figure 10.2, though, as remarked already, the ISI data reflect a rate of increase that is higher than that of the scientific population. To improve the model further, we would need data for the real scientific community, rather than as reflected by ISI so as to replace the uncertain assumptions displayed in table 10.3. Even if we had some confidence in setting the actual excess of recruits over terminators at a 7 percent rather than 10 percent level, we would still need an estimate for the rate of termination. The biological process of retirement and death of the scientific population would yield a rate of about 2 percent, but almost certainly an equal additional amount must result from transfer from active publication to teaching, administration, and other posts. With such an assumption (for which we can here adduce no empirical data) the model would have a recruitment rate of 11 percent, a termination rate of 4 percent, and a transience rate remaining at ca. 22 percent, which would imply that ⅔ of all new entrants were transient, and that only ⅓ of those making their first appearance(s) in publication will enter the community of continuants. The total newcomers being 33 percent, there remains 67 percent for the publishing continuants at the beginning of that year, and to these must be added a supernumerary 33 percent of continuants not publishing that year. The total body of continuants at the beginning of the year is therefore 100 percent and hence exactly the same size as the number of authors publishing, and the total scientific community is 1.33 times that number, including transients, recruits, publishing continuants, and nonpublishing continuants. The complete flow pattern for this improved model is shown in figure 10.3.

The agreement of this model with the empirical data is now excellent for the transients, the birth and death rates, and for the continuants who continue from one year to the next or have gaps in their publication records. It is still not quite adequate for the

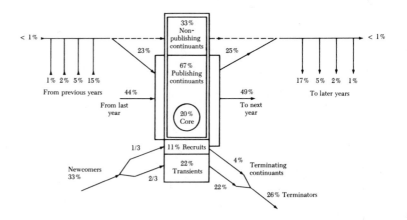

Figure 10.3. Improved model of flow of source authors

strong continuants who persist for several years together. Of the
96 authors recorded in 1964 there are 19 whose names occur in
all annual indexes through 1970. On the basis of a publishing
probability of $p = 0.67$ for a continuant population of 96, there
would be only 6 such people for the entire seven-year period; to
get 19 would require $p = 0.79$ for the whole continuant population. Alternatively one could say that there were 13 additional
authors so strongly continuant that for them $p = 1.0$ so they were
certain to publish in all years. From the data base we find the
actual proportion of authors who persist for at least an n-year run
relative to the continuants at the beginning of the period to be as
follows:

n	actual %	expected %
3	33	33
4	25	22
5	21	14
6	19	9
7	20	6

The expected proportions were all calculated on the basis of p
$= \frac{2}{3}$. It is apparent that the random probability model begins to
break down for runs of more than four years, and the constancy
of the subsequent actual percentages makes it natural to suppose
that there exists a hard core of highly persistent strong continuers
who amount to about 20 percent of the continuant population
(they must also be 20 percent of the annual author list) and who
publish without fail every year during their lifetime on the list.
This core, it should be noted, is a considerable fraction, 0.3, of
all the publishing continuants, and 0.45 of those continuing from
the previous year. No doubt the boundary between core and
noncore continuant publishers is not completely sharp, but some
graduation exists between those for whom the probability of pub-

lishing in any year is near unity and those for whom it is near the average value of ⅔; we shall retain the distinction as an aid to conceptualization.

With this refinement the model now accounts for all the empirical evidence from the data bank, and the components of the scientific community may now be summarily categorized as follows:

(a)   Transients who publish only during a single year and constitute 22 of those publishing this year, ⅔ of those newcomers to scientific publication during the year.

(b)   Recruits who begin publishing during the year and join the continuant population. They constitute 11 percent of those publishing, ⅓ of the newcomers.

(c)   Terminators who end their publishing during the year and thereby leave the continuant population. We have assumed they constitute about 4 percent of those publishing.

(d)   Core continuants who publish this year and indeed in every year for a long period. They amount to 20 percent of those publishing.

(e)   (Noncore) publishing continuants who publish this year and have a probability of ⅔ of publishing in any other year for a long period. They amount to 47 percent of those publishing, so that the total of core and noncore publishing continuants is 67 percent of those publishing.

(f)   Nonpublishing continuants who also have a probability of ⅔ of publishing in any year in question. They amount to 33 percent of those publishing, and this implies that the number of active researchers during the year is 1.33 times those actually publishing.

There is clearly a close relationship between the demographic structure that has now been elucidated and the distribution of productivity of scientific authors. Till now the data have been drawn from the mere appearance of an author's name in the series of annual indexes and not from the number of papers that he published in each year. The approximate average over the entire set of annual source indexes is that each author has his name on about two papers per year. Since, however, there are on the average also 2 authors on the byline of each paper, it follows that the total number of authors is equal to the total number of papers. Thus, although there is just one paper/author/year, there are about 2 authorships/author/year. Of these 2 authorships, about 1 is primary (i.e., the first author mentioned) and the other secondary—it is important to bear this in mind, for citations to papers are listed to the primary author only.

The demographic properties of transience and continuance are obviously to be associated with the lowest and highest rates of productivity, respectively. For transience, indeed, the publication is a one-shot event, and only formally do we associate with it an interval of a year. The training and research leading up to the event may take a much longer period. In a separate investigation we considered the publication records of authors who remained on the indexes for various spans of years. Those transients with a span of a single year produced 1.1 authorships during that year. Those continuants with spans of two, three, and four years produced 1.5, 1.73, and 2.0 authorships/year, and the obviously core continuants staying for five years had 3.7 authorships/year. Going even further to a nine-year span (extending to 1972) for the core, we found 4.3 authorships/year as the average for the group of 16 such authors. The continuants therefore have an average rate of production of about 2 authorships/year equal to the average of all the population, and the core continuants produce at about twice this average, balancing the transients with their minimal contribution of 1 authorship/year. In our small sample the most prolific authors had 14, 8, 6, 5, 5, and 4 authorships/author/year, respectively. In general the range of productivities seems remarkably low, running not much more than a factor of 2 above and below the average for everyone but the most prolific couple of members of the core group. It is well known that the work content of what constitutes a publishable paper varies quite a lot from field to field and even perhaps from country to country. Allowing for this, one might reasonably suppose that each of the three demographic groups have productivity factors that are not merely average but also typical. Transients appearing in a single year tend to produce a single authorship in that year. Continuants produce about 2 authorships (= one paper) and the core group about 4 authorships (= two papers) each year.

From this it follows that in any record extending over several years the numbers of papers produced by the various authors will depend more upon their lengths of stay at the publishing front than upon their rate of production while there. Unfortunately there is a basic difficulty in using such a long record to make deductions about the characteristics of the authors. The trouble is that at the beginning of such a record and, in the case of exponential growth, even more strongly at the end of the record, one is dealing with authors whose natural research lifetimes have been artificially cut. A cross-sectional study taken only over a given time range must contain a large number of authors whose careers began during that interval. For example, a ten-year study contains equal numbers of those beginning publication during the interval and those already present at the beginning.

Fortunately, however, there is now a great deal of evidence to show that such cross-sectional studies of productivity over reasonably long intervals of time follow rather simple laws with great regularity.[2] A first approximation is given by Lotka's Law which states that in any population the number of authors with exactly n authorships is proportional to $1/n^2$, and another form holding also for larger values of n is Price's Law which states that the number of authors with at least n authorships is proportional to $1/n(k + n)$, where k is a parameter of about 15 authorships/author/lifetime which marks a boundary between very high and normal production. The proportion of authors having a single authorship only may be calculated from these laws, Lotka's form giving 61 percent and Price's 53 percent. Both values are perceptively lower than the present demographic model in which we have settled on parameters such that ⅔ of all newcomers are transient, and hence there would be 67 percent of single authorship authors. The demographic model could be reconciled with Lotka's Law by keeping transients at 22 percent and taking the nontransient recruitment rate as 14 percent and with Price's Law by taking it as 20 percent. These latter figures are indeed exactly those of the second and first lines respectively of table 10.3 and are the limits of what could be maintained from the empirical data of the present study. They were modified, it will be remembered, only because it seemed excessive to have a 10 percent rather than a 7 percent per annum rate of increase and also excessive to allow a termination rate larger than 4 percent per annum. In any case, the changes introduced into the model by such variations of parameter are slight. They will have to be resolved in future work with larger samples of population which can be better specified now that a conceptual foundation has been laid for the chief demographic phenomena.

For those authors with more than a single authorship, the productivity laws seems to imply that the number of continuants with a given lifetime decreases with the length of that lifetime. Furthermore, the changeover in Price's Law at a parametric value of 15 authorships may well be identical with the demographic distinction that has been made between those continuants who are in the core group and those who are not. Clearly it should be possible to derive a lifetime expectancy function for paper publishers from the known productivity law. Just as the first year's publication record initiates one into authorship, and the second year's record selects the fraction who become continuants, so each successive year of publication reduces the ranks but makes it easier for those who succeed to continue. The final hurdle, that from continuant to core, must occur after something like five years of work and the production of some 15 authorships.

We suggest that these demographic stages that have been diagnosed correspond rather well to the social and institutional barriers that pervade every field and country in which there exist scientific and technological publication into the world corpus of common knowledge. It has already been remarked that the first barrier of securing the ability and permission to publish at all is that which is institutionalized in the United States and the United Kingdom as the Ph.D. degree. In other countries it may correspond to candidacy or some other postgraduate qualification. The second barrier is probably that of first acceptance into an academic, governmental, or industrial post in which at least part of the expected output is research front publication. The final barrier would seem to be that of the securing of tenure and seniority leading to a major lifetime commitment to research output and probably also the collaboration of junior continuants and transients.

It might at first be thought that the demographic structure that has been analyzed is a direct consequence of the institutionalization that we now have. Two factors militate against this interpretation. In the first place, it is quite clear that the productivity distribution of authors today is not essentially different from what it was for the seventeenth-century science published in the early volumes of the *Philosophical Transactions of the Royal Society of London*, long before the Ph.D. degree or the career scientist came into being. It seems therefore that the phenomena of transience and continuance must have occurred first, and the institutionalization followed. In the second place, the very form of the productivity distribution implies that behind the phenomenon of demographic stages separated by barriers lies a continuous process of people gradually falling away from active publication.

It follows from Price's Law that the fraction of authors proceeding from a total of at least n papers to the level of $2n$ papers is $0.5(1 - 1/(2 - k/n))$ where, as before, k is a parameter of value about 15 authorships. From this it follows that the transition frequency for authors to double their total of authorship begins at 47 percent for the first paper, falls to 33 percent at 15 authorships, and then decreases slowly and asymptotically to 25 percent for very prolific authors. Similarly, for a multiplication of output by 1.5 rather than a doubling, the transition probability begins at 67 percent, goes through 53 percent at 15, and ultimately decreases to 44 percent. For very large outputs this law ultimately breaks down because of the death of the author.

Another way of looking at the same consequences of this law is to say that at the beginning an author has an even chance ($p = \frac{1}{2}$) of multiplying his output by a factor of 1.89, by the time he has reached 15 authorships his even chance is for an extension by a factor of 1.56, and ultimately the factor is reduced to 1.41. In general, the pattern of the productivity law is such that what remains rather constant for any author is his chance of increasing his total output by a given ratio. Since we know already that authors differ not so much in rate of production as in duration of stay in research publication, these results may be interpreted as probabilities for the extension of publishing lifetime by the various factors. Again, what remains approximately constant is the chance that an author will continue for some multiple of his present span of years in research publication. Thus life expectancy at the research front is proportional to the time already spent there; the mortality rate falls steadily and hyperbolically from the very high level it begins with in infancy.

It is this steady force which produces both the productivity distribution and the demographic structure. For those beginning a publishing career the mortality is very high so that a majority of the publishing population falls into this category. With a minimal lifetime and a low rate of production it is easy to see that transients publish only a small minority of the totality of papers. In fact the 75 percent of the population who are least prolific account for only 25 percent of the output. At the other end of the scale are those authors who are so reinforced by successful publication that they have the smallest mortality, appearing for a long succession of years, and also having a high rate of production. This small core group of 20 percent of the continuants will probably produce more than half the total output.

All this is a result of the simple facts that success in scientific publication is extremely difficult to achieve and that success breeds further success—a good example of positive feedback or the Matthew Principle. The consequence of this situation is a strongly hyperbolic distribution of productivity, and the consequence of this is that the unit beginning of the distribution and its long tail both tend to behave as distinct entities which have here been identified as the transients and the core continuants. If this interpretation is correct, it seems so intrinsic that the institutionalization must be regarded as the way in which society has adjusted to the built-in pattern. For example, the barrier between first publication and second has its value by virtue of the difficulty scale of successive publication rather than because of the availability of socially useful jobs. With the same hyperbolic distribution now as in seventeenth-century England, it must be our society that has cut its various suits of institutional structure to fit the cloth of scientific productivity and demography.

11

Studies in Scientometrics, Part 2: The Relation Between Source Author and Cited Author Populations

In the preceding essay we used the techinque of taking a small alphabetic slice of a sequence of several annual source indexes of the *Science Citation Index* to elucidate the demographic structure of the scientific and technical publishing population. This technique is now extended to cover a series of citation indexes of the same volumes of *SCI* so as to determine the relation between the demographic groups of authors and the extent to which they are drawn on by the rest of the scientific community. It also reveals some information about the class of scientists who are cited but do not appear among the indexes of current authors.

First, it must be remarked that the annual citation indexes, which include all the references made in that year's papers, are necessarily much larger and more extensive than the corresponding source indexes. Usually they contain about 1.5 times as many names. Although only the name of the first author of the cited paper is recorded, the references cited each year are not limited chronologically. They not only extend back through the entire time span of all previous published journals, but also include all the nonjournal items cited. These include theses, printed books, patents, technical reports, and even informal publications that are specifically acknowledged by the citing author.

Coauthored by Suha Gürsey.

Even the most cursory examination is sufficient to reveal that a sequence of annual citation indexes behaves in very much the same way as the sequence of source indexes which has already been subjected to analysis. Two consecutive citation indexes overlap so that about half the names are carried over directly to the next year. An additional quota of names are carried over and reappear after a gap of a year or more. Each year there is a new supply of names that have never before been cited, and undoubtedly an investigation would reveal that there are also a set of names that never occur again. Thus we may suppose that there will exist classes of transients and continuants among the cited authors, too, and that, further, there will be authors cited so frequently that they must be counted as core continuants because of their appearance in every annual *Citation Index* over a considerable period. The parameters of probability and frequency governing this structure may be somewhat different for cited rather than source authors, but to a first approximation it would seem that the similarity is curiously close.

Using the same alphabetic sample as in the previous examination, a careful search was made for all names in three consecutive citation indexes, all names being edited into the same comprehensive list. Table 11.1 shows the empirical data derived for a span of source indexes 1964 through 1968, and citation indexes 1966 through 1968.

Five years of *Source Index* were regarded as sufficient to distinguish the transient and continuant classes among the authors, and the *Citation Index* was begun a couple of years later to pick up references to authors commencing publication at the beginning of the period. Even this short run is quite sufficient to show the main features. Basically our file listed 640 authors, of which 381 were in the five source indexes and 442 in the three citation indexes, with 183 names being common to both sets of index. It

Table 11.1

Cited authors	Not cited	66	67	68	66/7	66/8	67/8	66/7/8	Total
Source authors									
Not source	Not applicable	58	79	70	14	13	13	12	259
64	20		1		1		2	2	26
65	30	3					1	1	35
66	26	6	4	4		3	2	3	48
67	37	3	5	5	1	1	2	1	55
68	43	1	2	9		1		4	60
64/65	4						1	1	6
64/66	1	1					1	1	4
64/67	1			1	2		1		5
64/68							1	1	2
65/66	3	1		1		2		2	9
65/67	3	1					1		5
65/68				3					3
66/67	1	1		2		1	1	3	9
66/68	1			3		3		1	8
67/68	15			3			1	1	20
64/65/66	1						1		2
64/65/67	1						1	4	6
64/65/68									
64/66/67							1	1	2
64/66/68					1			2	3
64/67/68							1		1
65/66/67	1				1	1		1	4
65/66/68	1					1	2	1	5
65/67/68	2							2	4
66/67/68	3				2		2	7	14
64/65/66/67	1							3	4
64/65/66/68							1	2	3
64/65/67/68	1						1	2	4
64/66/67/68	1						1	3	5
65/66/67/68	1		1	1	1			3	7
64/65/66/67/68						1	2	19	22
Total	198	75	95	105	20	26	39	82	640

is worth noting that, as in the case of each index separately, about half the names in the *Source Index* set are carried forward into the *Citation Index* set. Indeed, the intersection between any two annual indexes, source or citation, not too far distant from each other in time, seems to be about half the smaller of the pair of indexes.

From table 11.1 it is clear that there are strong concentrations in two groups. A large number of authors appears in only a single *Source Index* and in none of the citation indexes. Another large group appears in all the source indexes and all the citation indexes. More particularly, of the transient authors who appear in only a single *Source Index*, 71 percent are uncited, a further 19 percent are transiently cited in a single index, and only 10 percent seem to be continuants in citation. These data are in good agreement with the simple probability model in which the work of a transient has about a 30 percent chance of being cited in any year; on that assumption there would be 70 percent authors uncited, 21 percent cited for just one year, 6 percent cited for two years, 2 percent for three years, and 1 percent for more than three years. For the core continuant authors who publish in all five years of record, 19 of the 22 are also cited in all three years and the other 3 are cited in two of the three years. It is worth noting that the 19 who are core continuants in both categories are also those who continue not merely for a five-year span but for the seven years recorded in the previous study.

For the noncore continuants of two, three and four years as a source, the incidence of appearance as a cited author is intermediate in behavior between that of transients and that of the core. For the group as a whole the behavior is precisely the opposite of the transients in that they have about a 70 percent chance of being cited, 30 percent uncited. As the number of years of continuance increases so does the chance of citation as shown in table 11.2.

The data probably overestimate the numbers of noncore continuants who are uncited, for, as can be seen from table 11.1, there are 15 cases of authors publishing in 1967/68 and 5 cases of those publishing in the same years and a previous year as well who were possibly uncited because their publications were still too recent. It follows that continuant source authors have a strong tendency to be continuants also in citation.

In addition to the demographic groups already discussed we have in the citation indexes a considerable group of people who were cited in one or more of the annual indexes but whose names do not appear among the source authors. Of the 259 names 80 percent are transiently cited in a single year, 15 percent appear in two years, and only 5 percent in all three citation indexes. We must, it seems, be dealing here with a mix of at least two different groups of authors. On the one hand, we have a group of former transient source authors and, on the other, the relatively small number of authors who were once continuants and are still fairly heavily cited even though they have retired from publication or died. The number of such persistently cited people not appearing in the source indexes amounts to 52, compared with an active continuant population of about 160 people in this period. This implies that about 75 percent of all the continuants who have ever been are still writing, but unquestionably the survival rate is

Table 11.2

Years as source	Number of years cited				Total
	0	1	2	3	
2	41	27	20	12	100%
3	22	10	24	44	100%
4	17	9	17	57	100%

higher than this, for some of the people cited must have published only in books or in journals not on the source list of ISI. The number who have ceased is thus less than 25 percent of the number of continuants who have ever been; this means that the number of continuants must have grown through more than a factor of 4—at least two doubling periods—in the interval of time it takes an average continuant to retire. With a growth rate of the usual value of 7-10 percent per annum this gives some fifteen to twenty years as the publishing life span of a continuant.

Such a life span seems rather too long for an average value between the few who spend a full working life at the research front and the many who spend one or two three-year terms in a junior appointment. The foregoing calculation needs modification if we suppose, as seems reasonable, that the 52 nonsource authors who appear in two or three citation indexes are not the only retired continuants. Suppose that there were, in fact, a retired continuant population of size C, each of which had a probability p of being in an annual *Citation Index*, and that successive appearances were quite independent. Then the number of cases of appearance in at least two such successive lists is given by Cp^2, and the number of cases of at least three appearances is Cp^3. Table 11.1 shows these numbers to be about 25 and 12 respectively; hence p is approximately ½ and C = 100. There are then about 100 terminated continuants instead of 52, and the number of them cited just in one index out of three is expected to be $100 (½)^3 = 12$. These 12 must be subtracted now from those who were previously counted among the transiently cited former transients.

We thus derive the new result that one component of former authors who are now cited consists of about 57 authors per year picked out of the very large number of former transient source authors and cited just once in a single year. The other component consists of about 100 former continuants who are no longer source authors but who have a chance of about ½ of being cited in any particular year. The number of terminated continuants being 100, and the number active being ca. 160 as before, the ratio of total born over total died is 2.6, and this implies that with a doubling period of 10 years we must take the publication lifetime of a continuant as about fourteen years. If one were to assume that all of these 259 cited nonsource authors were terminated continuants, and none of them former transients, the ratio would be 1.6 and the average lifetime a little less than seven years. If one were to suppose that there are even more former continuants who happen to be uncited in the three indexes, the ratio and therefore the lifetime would be correspondingly reduced. In order to make the lifetime as short as half a doubling period, say five years, we would have to suppose the existence of 384 terminated continuants, so that the 259 actually found in the three years of citation indexes would be only ⅔ of those actually existing.

After these considerations we are able to restate the categories of the scientific publishing community, amplifying the results previously found by the addition of the new citation data and the characteristics of the nonsource groups not before detected. The demographic groups now become as follows:

(a)   Transients whose names occur only in a single *Source Index*. They usually have but a single authorship (often secondary) in this year, and only a small fraction of these authorships, perhaps a quarter, are thereafter cited, usually just once in any year.

(b)   Noncore continuants whose names occur in several successive source indexes. They have 2 or more authorships per year, and have about a 70 percent chance of being cited in any year.

(c)   Core continuants whose names appear unfailingly in a long sequence of source indexes. They are about the most prolific 20 percent of all continuants, have 4 or more authorships per year, and also unfailingly appear in the citation indexes.

(d) Terminated continuants who formerly published for several years but remain cited with a probability of about 50 percent of appearing in any annual *Citation Index*.

(e) Former transients from all previous years who must be a population about twice as numerous as the continuant community existing at any particular time. Though they are therefore about 67 percent of the scientific population, they have published less than 25 percent of all papers, and these papers probably account for less than 10 percent, perhaps less than 5 percent of all citations.

(f) Newly recruited continuants who cannot in their first year be distinguished from transients.

Finally, as an additional bonus from the methodology of this investigation we are able to produce for this small alphabetic slice of the scientific population a pair of tabulations of the core groups who are continuant in the source and citation indexes.

Table 11.3 shows first the most-cited authors during the three-year period, together with their output of source authorship during the five-year period. Table 11.4 gives the most prolific authors with their citation record during the same period. There are 26 highly cited authors with an average of ten or more citations/year, including 2 who were not listed in the source indexes, and there are 25 prolific authors with an average of two or more papers/year, counting all primary and secondary authorships as a unit.

Common to both lists are some 15 authors so that the overlap is about 60 percent of either, a very striking correlation when one remembers that these are just the top sections of a pair of lists having 640 names each and only 147 in common. It would seem that the assumption that there exists a highly cited and highly prolific core group is quite justified. It suggests further that highly placed names not common to the two lists merit special investigation before letting these scores act as an evaluation. We have two clear cases of terminated continuants, several cases of the

Table 11.3

Authorships

26 most-cited authors	Citations	Primary	Secondary	Total papers	Years as source
*Palade, G.E.	1441	6	37	43	5
*Palay, S.L.	891	2	10	12	5
*Palatnik, L.S.	392	51	21	72	5
*Pais, A.	262	14	12	26	5
*Paivio, A.	250	14	15	29	5
Pake, G.E.	250	2	1	3	3
Pak, W.L.	135	8	—	8	5
*Palecek, E.	130	9	6	15	5
Pakula, R.	112	6	1	7	3
*Paladini, A.C.	108	—	20	20	4
Pal, M.K.	101	2	6	8	3
*Paladino, A.E.	87	12	3	15	5
Pakiser, L.C.	71	4	3	7	5
Pal, Y.	65	1	4	5	3
*Pal, L.	60	2	12	14	4
*Pakrashi, S.C.	59	15	1	16	5
Pal, S.	51	7	2	9	4
Palais, R.S.	48	2	—	2	2
*Pais, M.	44	7	5	12	4
*Pala, G.	39	15	10	25	5
Paldino, R.L.	35	3	3	6	5
*Pakvasa, S.	32	9	9	18	5
*Palaic, D.	32	8	2	10	5
*Pakkenberg, H.	30	12	8	20	5
add to these, non-sources:					
Palache, C.	189	0	0	0	0
Pal, B.P.	40	0	0	0	0

*Authors common to both lists (see table 11.4).

prolific production of apparently little cited work, and several cases of highly cited authors of rather few papers. In the last class, however, it must be remembered that some of these highly cited authors might be in the course of termination and are producing

Table 11.4

| | | Authorships | | |
25 most prolific authors	Citations	Primary	Secondary	Total papers
*Palatnik, L.S.	392	51	21	72
*Palade, G.E.	1441	6	37	43
*Paivio, A.	250	14	15	29
*Pais, A.	262	14	12	26
*Pala, G.	39	15	10	25
*Paladini, A.C.	106	—	20	20
*Pakkenberg, H.	30	12	8	20
Paknikar, S.K.	8	—	19	19
*Pakvasa, S.	32	9	9	18
Pakhomov, V.I.	18	6	10	16
*Pakrashi, S.C.	59	15	1	16
*Palecek, E.	130	9	6	15
*Paladino, A.E.	87	12	3	15
Pak, C.Y.C.	20	8	6	14
*Pal, L.	60	2	12	14
Pal, A.K.	15	7	7	14
*Pais, M.	44	7	5	12
*Palay, S.L.	891	2	10	12
Palacek, J.	3	7	5	12
Pakalns, P.	28	11	—	11
Pakala, W.E.	6	5	7	12
*Palaic, D.	32	8	2	10
Palamidessi, G.	4	4	6	10
Palacios, O.	5	4	6	10
Pak, M.S.	0	—	10	10

*Authors common to both lists (see table 11.3).

much less than in the period for which they are cited. In only one case is there something as flagrant as an author who has published ten papers but received no citations.

12

Of Sealing Wax and String:
A Philosophy of the Experimenter's Craft and
Its Role in the Genesis of High Technology

In the 1920s, it was hard to find a piece of research physics equipment in British laboratories that was not stuck all over with red Bank-of-England sealing wax because this was the best cement available for holding a vacuum. A little later, when apparatus had to be demountable and scientists had to be able to break and regain X ray vacuums quickly, the cement of choice became plasticine. During the golden age of experimental physics early in this century all progress seemed to depend on a band of ingenious craftsmen, with brains in their fingertips, and a vast repertoire of little-known properties of materials and other tricks of the trade. It is these that made all the difference in what could or could not be done in a laboratory, and that, to a large extent, determined what was discovered.

The phenomenon is not confined to physics, to any particular country, or even to the present century. At one point in the mid–nineteenth century, knowledge of new synthetic dyestuffs dominated biological research which could be used to stain substances and thereby reveal new structures in tissues under the microscope. For a long time, this gimmick, rather than the optics of the microscope, was the key to new scientific exploration. Similarly, tricks with polarized light were applied to all sorts of fields,

Adapted by Robert K. Merton from the George Sarton Memorial Lecture given at the American Association for the Advancement of Science in 1983.

and were often taught as elegant philosophical toys and games; Albert Michelson used polarized light to discover the constancy of the velocity of light.

One has only to read such masters as Galileo, Newton, Maxwell, and Einstein, and inventors such as Edison, to realize how each, in their very different styles, set great store by and derived great benefit from such miscellaneous craft information that could be used to do things in science. Be it noted also that great masters of this province were not by any means always great cognitive contributors to science. Many were almost anonymous and un-sung lab assistants, such as Lord Rutherford's man, George Crowe, or J. J. Thomson's aides, Ebeneezer Everett and W. G. Pye. These three assistants went on to found the Cambridge Instrument Company, one of the first high-technology electronic companies of Britain. Thomson and Rutherford were genius ex-perimenters who happened to be rather clumsy, and their assist-ants were crucial to progress. In fact, much of the apparatus used in the Cavendish Laboratory at Cambridge, where both men worked, was held together by sealing wax and string, not out of poverty, but because the genius experimenters had a dozen pairs of clever hands feverishly tearing down and rebuilding the ap-paratus as the techniques were pushed in new directions and improvisations.

The flavor and tradition of this sort of experimentation are markedly different from, and perhaps even in conflict with, the view of the role of experiment in science that one might derive from standard texts in the philosophy of science. This view is found in those banal texts that preach on The Scientific Method, and it is implicit in the writings of that majority of my colleagues in the history of science who approach science as theoreticians and have little feeling for bench science. Interestingly enough, a new breed of historian, laboring in what they call the "anthropology of science," has been looking at the behavior of scientists in laboratories. The findings are suprisingly different from what one would expect, if one were dealing with an exercise in intellectual history.

The standard view, which I ask you now to reject as being rare in history and not at all the essence of the scientific enterprise, is that the scientist creates hypotheses and theories and sends them out to be tested and tried, confirmed or (with Popper) falsified by making a trial of the prescribed "experiment."

What actually goes on in laboratories is of a different nature. Since the seventeenth century, and perhaps even earlier, experiment has more often meant "experience," as various techniques, like sealing wax and string, are used in every which way in the hope that the finding out will reveal facts of nature that fall outside the range of what was known before. The procedure is far from being cut and dried, and the theoreticians and experimenters are far from being in the master/servant relationship in which they are usually cast. In many societies there is a clear social class difference, a put-down of the experimentalist as a "mere engineer" doing the bidding of the creative intellects responsible for the prized cognitive advances. On the contrary, skilled experimenters are masters of a very peculiar and crucially important sort of technology. Their work is close to being the core of high technology and represents a tradition that is rather autonomous, arising not from the cognitive structure of science, but from other adventitious technologies devised for quite different purposes. All I am asking for is a more reasonable balance which recognizes that from time to time, perhaps even predominantly, experimental craft and art provide an exogenous force that moves science rather than does its bidding.

Another remarkably widespread wrong idea that has afflicted generations of science policy students holds that science can in some mysterious way be applied to make technology. Quite commonly, is is said that there is a great chain of being that runs from basic science to applied science and thence to development in a natural and orderly progression that takes one from the core of science to technology. Historically, we have almost no examples of an increase in understanding being applied to make new advances in technical competence, but we have many cases of advances in technology being puzzled out by theoreticians and resulting in the advancement of knowledge. It is not just a clever historical aphorism, but a general truth, that "thermodynamics owes much more to the steam engine than ever the steam engine owed to thermodynamics." Again and again, we find new techniques and technologies when one starts by knowing and controlling rather well the know-how without understanding the know-why. We often (but not always) eventually understand how the technique works, and this then leads to modifications and improvements, giving the impression that science and technology run hand in hand. But historically the arrow of causality is largely from the technology to the science.

The simple truth is that if one wishes to do something to something, what one uses is a technique rather than an idea. A hackneyed example of this is the application of Maxwell's electromagnectic theory to the invention of radiotelegraphy and then all the technologies of radio and TV broadcasting. His theory was a tremendous unifying concept that explained the nature of light and suggested that one could produce similar wave radiation electrically. The trick was not knowing that it might be done but finding out ways to generate and detect such waves.

In particular, although basic and applied science may use the same methods and techniques, applied is not an application of basic. When one turns a spectroscope on the stars it is basic research, and when one turns it on a steel furnance it is applied research. Development is something else again. Strictly speaking, it has nothing to do with the growth of science or technology since its labor force is that of production rather than people in whom we have invested an expensive scientific and technical education to put them at the frontier of knowledge and capabilities.

If experiment were only the handmaiden (or some other nurturing sort of lackey) of the makers of theory, then the history of technology would be the social history of business enterprise, and invention would be a set of footnotes appended to the history of science. Correlatively, if all applied science were produced by reflection or an autonomous world of industry, a lot of the history of science would be footnotes appended to a history of technology. Neither of these, of course, is anywhere near the truth. The truth ought to be a dialectical interaction of science and technology, but alas, we don't get this either. Our parascientific professions happen to have been professionalized quite separately so that most historians of science get a certificate of ignorance in the history of technology and vice versa. In order to write a concurrent history of science and technology that is more than a mere chronology, we must be concerned with the historical causality of it all. Why do the events happen when they do and the way they do? One key to understanding lies in this region of the techniques of experiment that have dominated science since the seventeenth century.

Let us now turn to the historical development of this craft of experimentation and scientific instruments that occupy pride of place in the story. Again we have a long-standing misapprehension to dispel, arising from the supposition that the experimenter is doing the bidding of the real (theoretical) scientists. It is held that instruments began as tools of measurement, starting in astronomy and extending to the related crafts of time-keeping, navigation, surveying, and gunnery. This is far from the truth. In ancient astronomy one needed only qualitative observations of crucial cases separated by long time intervals to obtain parameters accurate to several places of sexagesimals. Ptolemy was a master of this, and used such nonmeasurements most judiciously. Anybody who used instruments for pretelescopic naked-eye astronomy must have come to the conclusion that they were inferior to visual estimation since a reasonably sized instrument with degree divisions about a millimeter wide cannot master heavenly bodies better than one degree or so. But the naked eye, perceiving the width of the rather big sun or moon (each being about ½° wide), could make estimates four or five times more precise than that.

Why then does Ptolemy describe elaborate instruments? The reason is twofold. First there is the difficulty that theory of planetary motion, which is the main burden of ancient astronomy, proceeds from measurements of latitude and longitude relative to the invisible ecliptic rather than from altitude, azimuth, and proximity to the moon and bright stars. Converting from the one set of coordinates to the other is very tedious, and it was nice to know that with an instrument you could measure what you wanted to measure directly. Second, it happens that the best crafted instruments of antiquity right through the Middle Ages were not instruments of observation at all. The astrolabe, for example, was a plane simulation of spherical astronomy. It might have been used for calculation, and I suppose it was often suggested but almost never used as a divided circle for observation. I suppose it was popularized to show the virtues of useless but beautiful mathematics. Actually, it was probably made and acquired for much the same reason as we have globes in an elegant and learned library, as an embodiment of theory. Astolabes were not even used for practical pedagogy but to symbolize the possession of a theory.

Through the ages, many other varieties of instruments served a similar purpose. Ancient sundials were not so much for telling the time in a clockless world as for symbolizing the inexorable circles of the sun's motion through the year and the day. Surveying devices popularized geometry and trigonometry, but one should not believe that estimating the height of inaccessible towers on the other side of the river was ever a vital need of society. Good navigators never trusted their few instruments, nor, indeed, should they have. An eyeball estimate of the height and direction of the polestar was all one needed, and for this one needs no instruments. Compasses are useless until one has another global knowledge of magnetic declinations to know which direction the needle should point (it can be as much as 25° east or west of north).

In short, I do not believe that there was serious measurement with any instruments before the sixteenth century, but there did emerge a high craft of making models and simulations such as astrolabes, armillary spheres, globes, and sundials. In the sixteenth century, gunnery instruments came into vogue, but one only has to look at them to know that they were symbols, devices to let one know the master gunner was an educated man, rather than working devices used in the heat of battle.

Of course, instruments did not remain only symbolic. Two major changes occurred in the history of experimentation that brought about revolutionary transformation in the nature of science: Galileo's use of the telescope in 1609, and the Galvani and Volta discovery of current electricity about 1800. We shall examine both these well-known stories to identify the patterns of historical causality that are archetypal for science and technology and still dominate the way in which both actually evolve—a way very different from that popular in circles where the philosophy of science tells the way in which it *ought* to work.

Galileo made his first telescopic observations and published them in a little book, *The Heavenly Messenger*, in 1610. It made his reputation, achieved the so-called Copernican Revolution, and popularized what was known for the next century as "the new philosophy." Ironically, the whole chain of events occurred because of technological availability in a wholly unrelated area. The parent technology was that of eyeglasses, a craft trade that grew with the manuscript copyists of the thirteenth-century age of the great cathedrals and monastic institutions. It flourished mightily when printing moved books and the craftsmen to the cities and proliferated them. In the late sixteenth century the glass lathe was introduced, enabling several lenses to be ground at once and also to produce, as objects of curiosity, powerful thick concave lenses.

Thin concave lenses had been used for short sight for more than a century, but the thick concave lenses were now sold only to people caught up with the world of painting and illusions of perspective who used them as "perspective glasses" to see the world in a virtual image microcosm. It is only when this new sort of lens became available that there was the possibility of seeing an interesting effect by combining two lenses. We now know that many different things can be done with a pair of lenses, but the Keplerian telescope and the microscope both demand that everything be in nearly perfect focus before you see any more than a blur. With the Galilean type of telescope, as soon as you hold a powerful concave lens to the eye and a simple weak convex lens at arm's length, the clock in the church tower jumps out toward you. Many artisans around the world enjoyed the interesting illusion of distant objects brought near, but it was a pair of lens grinders from Middleburg in the Low Countries who first decided to market the device as a military invention. The reason was not that it was well suited as a spy device, but simply because that is where the money was. When the telescope was used militarily, centuries later, it was not for spying but for signaling.

Galileo was consulted by Venetian authorities about the new

lens because he was an uncommonly aggressive would-be consultant, part of the first generation of university professors who had to get their fees from students and patrons rather than from the establishment. He seems to have duplicated the invention and used it with little difficulty despite its tiny field of view. The decisive and traumatic event for Galileo must have been when he looked at the moon for the second time, a few days after the first view, found that the shadows of the mountains had shifted, and easily estimated that they must be about the same height as the mountains of the earth. That the moon has mountains and seas suddenly changed from an illusion into a fact. It was, moreover, a fact of nature that nobody had known before.

The enormity of the discovery must have been apparent immediately, and it changed for all time the nature of scientific scholarship, not just the career of Galileo or the theories of astronomy. For the first time in history, a person had made a discovery not available to other people and by a process that did not involve deeper and clever introspection. Galileo had discovered what was effectively a method of artificial revelation that promised to enlarge what was to be explained by science. It was the method rather than the accident of the particular results that created furious opposition from the Church conservatives, embattled ideologically, politically, and economically in facing the Reformation. In modern times people have felt the same way about the claim made by some that ingesting a little lysurgic acid could reveal new truths about philosophy beyond the reach of unaided mortals. At all events, it seems clear that the previously unknowable facts about the satellites of Jupiter, the phases of Venus (which showed it to be not self-luminous, but lit by the sun), the enormous number of dim stars, and the like quite changed the picture of the universe to be studied, making it Copernican not Ptolemaic. In terms of planetary kinematics alone, sans telescope, no worthy astronomer would have made the switch to a model that seemed rather more complex and perhaps less accurate than the old model.

The telescope became a craze not just because of the particular discoveries. It provided a chance to "tune in and turn on" to the New Philosophy of using instruments to find out things beyond the reach of the natural senses, and not deducible by mere brain power. Every available technology was mobilized to develop more instruments with which unsuspected facts of nature could be discovered. Once one had the telescope it was natural to study optics and find out why the device worked. From that study other types of telescope emerged and so did the microscope. Suddenly all manner of virtuosi were trying to dream up new ways of modifying lab instruments and of imbuing them with the magic of the telescope. Modifications of the pumps for firefighting and mining yielded the vacuum pump and created an awareness of air as a gas rather than as an all-pervasive medium. Thermoscopes and thermometers created a new world in which one thought more clearly about heat, knowing that neither pepper nor passion was really hot.

In short, the dominant force of the process we know as the Scientific Revolution was the use of a series of instruments of revelation that expanded the explicandum of science in many and almost fortuitous directions. Of course, there was more to it than that. There were all the powerful forces of mathematical analysis that Newton brought to bear on dynamical astronomy and mechanics which brought about the first great unification in modern science. Then there were the social forces binding the amateurs together into their first scientific academies, providing them with the scientific journals that led scholarship to advance with exponential rapidity in small step-by-step advances that moved far more quickly than was possible for an unaided set of humans writing lifetime books. But predominantly it was the instruments, not any special logic of Francis Bacon, that gave rise to the philosophy, and those instruments came out of technologies that owed nothing to motivation from physics, owed nothing to problem-solving activity. If you did not know about the technological opportunities that created the new science, you would understandably think that it all happened by people putting on some sort of new thinking cap. It is precisely that error which has been made by such eminent historians as Butterfield and Kuhn. The changes of paradigm that accompany great and revolutionary changes may sometimes be caused by inspired thought, but much more commonly they seem due to the application of technology to science. Furthermore, the instruments were not passive tools created to do jobs-to-order, and the jobs done were other than exact measurement. The only genuine measurement used to found a theory outside of astronomy that I know was the Boyle-Hooke determination of the elasticity of air; moreover, that had little direct effect and did not give rise to any tradition of measurement in science until about a century later.

The forces of the Scientific Revolution were pretty well spent by the end of the seventeenth century, and all was relatively quiet again until the end of the eighteenth century, when yet another revolutionary advance in the technology of science rocked science. It had repercussions all over science and technology and created entire sets of new methodologies from which a world of what we call high technologies was to emerge.

The key in the long chain was the discovery of current electricity. The anatomist and physiologist Luigi Galvani had been working on the very hot research front of using an electrostatic generator to give frogs a shock, making their muscles contract. It was supposed that the electric fluid passing into a frog's nerves was stimulating its vital fluids, to produce an action which was the very stuff of life, and that one might thereby discover the basic secret of the life force. Galvani noticed that the muscles would twitch even when the electric machine was disconnected, and Volta found that you did not really need the frog, either. A pair of metal discs sandwiching something moist and acidic produced the taste of electricity on the tongue, and a pile of such pairs would give a plentiful supply of the electric fluid. This fluid or current was immediately recognized as a powerful chemical agent capable of causing the first new chemical effects since fire and water. Electric fluid could decompose water into gases, and could plate metals. In a remarkably short time, many new chemical elements were discovered through electrolysis. Once people learned about such active substances as sodium and potassium, a whole lot of new chemistry was up for grabs. The chemical revolution resulted much more from the technique of the electric battery than from the careful measurements or new theories of Lavoisier.

Within less than a generation, chemistry changed from a deskful of alchemical supplies and methods into the laboratories of scientists like Davy and Faraday and Liebig. The chemical revolution so produced was not just a cognitive advance; it also led to immediate and spectacular change that gave us high technology and transformed the lives of all mankind. In the first generation of chemists who could analyze and synthesize at will, there was Liebig's agricultural chemistry with its fertilizers that produced plentiful food from barren land, artificial dyestuffs that gave new life to the textile revolution, and turned the everyday world from drab into color. Not long after that, there emerged antiseptics and anesthetics that led to a new technology of surgery. While all this was going on, lab scientists continued to study the nature of electricity. It was soon enlarged from its role as a chemical juice to become recognized in its almost mystical relationship with magnetism, and to change into a new sort of energy that led to the age of Edison. Such a capsule history can only hint at the forces that produced more than a century of better living through chemistry and through electricity. And it all happened from a lab experiment to find the secrets of life in the back legs of a frog.

An interesting sidelight on this phase of scientific experimentation is that for a brief while the research front was dominated by a lot of small pieces of electrical and magnetic apparatus that could be connected by wires in a variety of ways, and by the test tubes and reagents of wet chemistry with which all manner of experiments could be done. It was in this stage, I suppose, that the idea became current that it was out of such experimental trials that the work was done. It was precisely during this period that pioneers such as William Whewell were writing their histories and philosophies of science that canonized these special procedures as The Scientific Method. Science had changed from a new philosophy of instrumental revelation to the methodical testing of a rich store of hypotheses in electricity and chemistry. What happened also with the mushrooming of new techniques in electromagnetism, and new methods and substances in chemistry, is that each technique became a candidate for exploitation as a technology, possibly as lucrative as fertilizers and dyestuffs, the electric light and the motor. It is from this period then that the public began to identify science with technology, and the method of science as the testing of theories.

In the 1830s chemical laboratories entered into the universities as a necessary component of instruction. In part this was for actual education, and in part it was to train the new working classes of the industrial revolution. In the 1870s, the first physics laboratories emerged as part of the same process. Students were taught the art of precise measurement on such instruments as the Kew magnetometer—again supporting the widespread idea that measurement was the heart of empirical procedure. Around 1896, with the unexpected new technique of X rays and the phenomenon of radioactivity, the world of the physical sciences broke out again from its established confines, and the laboratories began a new quest for sophisticated new effects that would reveal the universe in different terms.

It is during the ensuing period that the prime experimental techniques began to involve the preparation and maintenance of a better and better vacuum, and it is this that filled the labs with the characteristic sealing wax and string. During the next several decades what scientists hoped for in all the routine experimentation that constitutes what Kuhn calls Normal Science was the discovery of a hitherto unknown technique, preferably one that was aesthetically pretty as well as interesting and revealing. One did not have to understand a new technique in order to find it peculiarly useful. The cloud chamber that enables one to visualize the tracts of atomic particles started as an attempt by the mountaineering buff C.T.R. Wilson to make artificial mountaintop clouds in the laboratory where Rutherford happened to be doing his radioactivity experiments. Rutherford, who employed George Crowe as a lab assistant, discovered induced radioactivity while doing a dull series of measurements designed to test out his ideas about the transparencies of several gases to alpha particles. Each effect, unpredicted, provided a technique for doing several things that led to new phenomena and even to further techniques. Not all pretty effects were recognized for the real utility they had. I remember the principle of xerography being shown as an instructive parlor trick to elementary physics students long before its vast industrial exploitation.

It would be easy to extend this litany of technological devices applied to science right up to the present. Any effect or phenomenon, such as the Edison effect, Cerenkov radiation, and the creep of liquid helium, might be just the thing to measure or reveal something we did not know before. Such experimentation is a sort of fishing expedition because you never quite know what you will catch, or what will happen. You always hope for the unexpected. It cannot be planned with an eye toward any particular objective, though of course it is common to cite some goal as a necessary condition for getting funding. If a technique can be exploited in a new way, we must push—even though we know not where it will take us. It also seems clear that we never quite know whether the technical trick that is so close to the nexus of excitement in the laboratory may also be marketable by some bright entrepreneur. I remember Patrick Haggerty, founder of Texas Instruments, telling me of his amazement that a gimmick he had dreamed up to exploit the first mass production of transistors ran away with the market. Nobody predicted that the transistor radio was more than a quick Christmas novelty. Nobody realized it would sell all over the world and would open up whole parts of the globe to modern modes of communication.

Since the advent of the technologies of radio, computers, and accelerators, there has been a strong trend from small unitary pieces of apparatus back to specially built instruments which are manufactured and used as bought. Sometimes, as with the big accelerators and radio telescopes, the instrument may be a large engineering construction and constitute a very large institute in and of itself. At the other end of the scale, it is common to find methods and techniques that use no more hardware than can be found in any elementary laboratory. In that case, one only has to be told or shown the "trick" to be able to repeat it at will. Many of the most-cited scientific papers of all time, such as Lowry's neat method for protein analysis, are exactly of this variety.

Technology transfer seems to be of especial strategic importance in all the techniques, tricks, and effects which are the repertoire of the well-trained and ingenious experimentalist and also of the key inventors of high-technology industry. Often one cannot explain why the techniques work or foresee just how they will be useful. A good case in point is that of the high-temperature gradient furnace which was developed very carefully and cunningly to make artificial gems. It was brought to perfection when it was discovered that the furnace was particularly effective for making rubies just at the time that rubies for minute bearings became an important strategic material in World War II. Bell Labs then used the method to grow single crystals of pure metals to see if the knotty problems of metallic conduction could be solved in the absence of impurities and crystal boundaries. It did not help, so they were led on from old acquaintance with metal whiskers and the crystal rectifiers of early radio to try the semi-metallic elements. It is this that gave them the transistor effect. As luck would have it, it was also just the time that the transformation of the calculator/computer from gearwheels to vacuum tubes had finally bogged down into impossibility (because the tubes kept blowing), so they were waiting for something new.

Although history testifies that these tricks, techniques, and instruments are of crucial importance to both science and technology, they are badly served by our social arrangements for promoting both. In science, whenever money is short, we prefer to spend it on people rather than hardware. In social standing, the technical experts and the people with brains in their fingertips are regarded as servants of the cognitive people with theoretical training. In technology, the instrument industry is quite insignificant compared with giants like automobiles and steel, even though it might be the point of orgin for giant inventions that could set off whole new industries. There are nations like the Soviet Union, most enthusiastic for all new science, but where kids do not have good scientific junk to play with so as to train the brains in their fingertips. Every year in America we spend a little less on hardware and on the relatively undirected experimental play with it. For 400 years we have been transforming the world by applying technology to science and thereby winning new techniques as well as new understanding. Surely it is time to recognize what a powerful force the experimental craft and technology have been, and to incorporate the unmistakable lessons of history into our current policies for science and technology. This would also deepen our understanding of that history itself.

13
The Citation Cycle

This essay had its origin in a chart, long resident on my office wall. Frequently revised, corrected, and redrawn, it seems to have taken on a life of its own and some utility to colleagues and students who acquired copies at various stages during more than ten years. The basic idea is to exhibit an interlocking metabolic complex of bibliometric (and scientometric) parameters in a comprehensive and integrated structure after the manner of the nitrogen cycle and other such paraphernalia beloved of organic chemists and ecologists. The data for this cycle are those drawn from the largest collection we have of machine-handled and automatically counted bibliographic items—the *Science Citation Index (SCI)* which has been published by the Institute for Scientific Information (ISI) since its foundation by Eugene Garfield in 1961. In biblio- and sciento-metrics it is often fatal and invariably debilitating to do your own counting. Beyond the tedious work and expense there is a hidden danger that one might well falsify the investigaton by artifacts of definition and selection, so it is far better to use unobtrusive indicators produced by people who didn't know you were going to use them thus.[1] Much of my research in this area has been fed by a steady diet from the cutting room floor of printouts produced by ISI partly in their direct function of producing what is not only primarily a bibliographical aid but also *the* chief bibliographical service for scientists. The other part has been composed of special printouts generated by their admirable curiosity about their own processes, for which I am truly grateful.

An incidental advantage of this parasitic nourishment of my work is that the data, most of which are now conveniently published on an annual basis in the preambles to the *Social Science Citation Index*, the *Science Citation Index*, and the *Who Is Publishing in Science* volumes, cover a large range of that which is implicated in the available corpus of both bibliometric and scientometric research theories. The citation cycle therefore embodies many of the elements of theory which are treated in the scholarly literature in our fields,[2] and it thus provides a sort of overview and coherent conspectus of a framework for the theories.

A tour of the citation cycle begins (see figure 13.1), as does the formation of a citation index, with the selection of the source journals and the items (usually research papers, but the more general term is useful) which are contained in them. The *selection* of journals is crucial to the success of a citation index because it is a strategy quite different from the usual librarian's striving for completeness. Though one may well start from an attempt to include all significant journals within some definition from all countries and all fields as sources, the ultimate test is provided as feedback from the journals which are cited by such sources. For many years the list of cited journals has provided a higher criticism of which journals to accept and which to reject as sources. Some journals may be so esoteric or so local that the citations they receive are from themselves. Others may have purposes of news and current awareness rather than the communication of citable knowledge and be for that reason almost uncited even by themselves. Then again some of the most cited journals may be extinct or living under a new name, or they may use the archaic practice of incorporating references in the body of the text where it is too expensive to employ keypunchers to excavate them.

If ISI chose its ca. 2700 source journals at random, they would be only about 6.7 percent of the (maybe) 40,000 scientific and

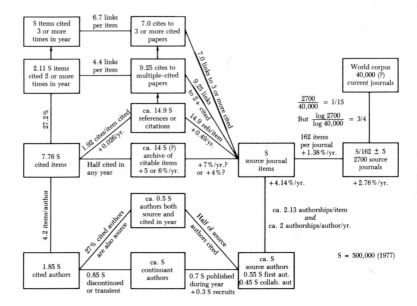

Figure 13.1.

technical journals extant in the world, and hence they would contain only a comparable fraction of the current source literature. If ISI were perfectly successful, as no doubt they are not quite, in skimming only cream, they would get as sources just those source journals which were the most cited. In that case one can apply the powerful principle of Bradford's approximation to the distribution law of cumulative advantage in journals;[3] cumulating citations from the most-cited journals downward, the total of citations is proportional to the logarithm of the number of journals included. This is much more realistic, and it has the advantage, as it should, that the result is not at all sensitive to the count of all the world's journals—a ballpark estimate will serve. The result of this estimate is that the *SCI* now includes log2700/log40,000 = 0.75 of all cited papers. Thus although it is derived from only ¹⁄₁₅ of the source papers, it includes ¾ of the cited literature. As a corollary we may now claim that if the data in our citation cycle are multiplied by ⅓ they will give the world data for the cited corpus.

The 2700 source journals did not come all at once. The first few numbers of annual publications were based on about 600 journals, and then in 1964–67 there was a period of expansion and revision (see fig. 13.2). From 1969 onward the number of journals has been expanding at an exponential growth of 2.76 percent a year (derived from a regression of the logarithm of the number). This is much smaller than estimates of the world growth of scientific literature, 6–7 percent a year, so we are dealing with a relatively unchanging core of journals. The number of source articles in these journals is now about 500,000, and it has been growing since 1969 at a rate of 4.14 percent a year; it follows that on the average the journals have become slightly fatter at a rate of 1.38 percent a year. Apart from this slow change we can say that although there is considerable variation in size between journals, on the average

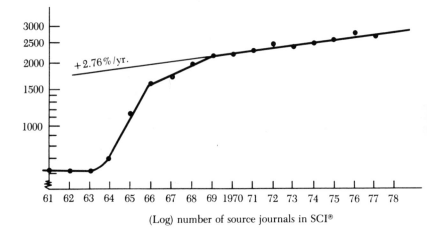

(Log) number of source journals in SCI®

Figure 13.2.

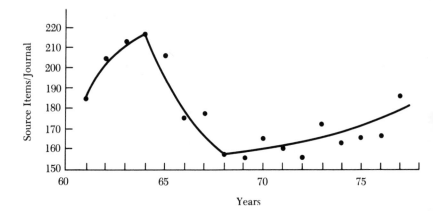

Figure 13.3.

each contains about 162 ± 5 source items/year (see fig. 13.3). Note the sharp drop in average size during the 1964–68 expansion as smaller journals are added. This is an interesting size, for it is equal in magnitude to an average invisible college of co-workers, usually 100–200 people, each writing about one paper a year in any of the major subdisciplines into which science is divided.[4] One might conjecture that a journal comes into being to serve such an internally communicating group of researchers, and then in the normal process of aging as the invisible college grows and produces new groups by fission some of the journals survive as media for aggregates of the living subfields.

The next stage of the tour of the citation cycle connects the number of source items (we shall designate this as S henceforth) with the authors of those items. In the dim distant past of science, from the late seventeenth century when scientific journals began until about World War I, when collaborative authorship was a rather rare event, the norm was that an active researcher produced about one scientific paper per year.[5] Professionals tend to have a discretionary period in their life-style which runs on an annual basis, and in harmony with our annual reporting of activity the normal life cycle of a project tends to be adjusted to this calendar cycle. What has happened since that period, and with great rapidity in the time since World War II, is that scientific authors collaborate increasingly so that in most scientific fields there is an average of two names or a little more per paper. What is happening is that the developing entrepreneurial tradition of channeling research support funding through a principal investigator in effect permits that person to purchase subsidiary authors. The result is that the number of authors per paper has become a rather good indicator of the extent of grant support in the field. Cancer and heart disease research is highly collaborative, pure mathematics much less so; it may be that in fields that need big team work the grants have to run high, but the effect may just as well be the other way round in causality. At all events, even though it now takes two authors to produce a paper, the output in papers has stayed constant, for now instead of each author getting out one paper per year, the team of two on the average produces two papers per year. The result is that the number of source authors is also S, and to be more precise there will be among them 0.55S primary authors and 0.45S secondary authors. Also, to be a little more precise, there are now 2.13 authorships associated with each paper, across all fields. It should be noted that this group of statistics varies quite a lot from field to field, perhaps even from country to country. There are some fields like systematic taxonomy in natural history, or certain parts of organic chemistry where a paper may correspond to only a few weeks' work, and there are fields like astrophysics where an ordinary research contribution may be of two years' duration or longer to make a single paper—such goes the size of atoms of knowledge in various disciplines.

For the next stage in the tour we enter the domain of citations. Each paper includes a list of articles to which it refers. The references are usually at the end of the paper or are footnotes on the page, and in the formation of the *SCI* these are keypunched into the computer record to be sorted into a citation index, alphabetic by cited author. Although the source items include everything from those totally devoid of references, e.g., news items and pontificating remarks, to those with hundreds or thousands of references in a bibliography, on the average there are about 14 references from each of the source items. In fact, cumulative advantage theory shows that what is really happening is not to be thought of as the new papers making reference back to the old; it is the old papers that are throwing off citations every year and thereby making occasion for the new literature. At all events, the average number of references in a paper is determined by the size of the available archive of literature in that field. Indeed the number of references per paper must be a small constant (less than one) plus the natural logarithm of the size of the archive. The natural logarithum of one million is about 14, and that is why the number of references is what it is.

For the *Social Science Citation Index (SSCI)* the corresponding number is about 11 references per paper, which is what would happen for an archive of about 60,000 papers in each field. Both in the *SCI* and in the *SSCI*, the number of references per paper has been increasing as the archive has grown (table 13.1). For the *SCI* there has been an increase of just less than half a reference per year (0.49), and for the *SSCI* the value is 0.62 per year. For the *SCI* the relative growth in number of references is about 3.5 percent a year and for the *SSCI* about 5.5 percent, corresponding to rates of growth of the archive at these values. Though both are lower than the traditional 7 percent per year growth rate of all scientific literature that we used to assume, they match the growth rate of source articles reasonably well. One must suppose that the ISI corpus is now growing at little more than half the historic long-term growth rate of the literature in the past century or so.

The references back from the source papers fall upon the available archive of papers already published. As we shall see, only about half of this archive is cited at all in any particular year, but of those papers that are cited a large majority, 72.8 percent, are cited once only. Of the remaining papers about half are cited just twice, and though the number of papers falls off very rapidly at about the inverse cube of the number of citations, there are still $\frac{1}{400}$ of the items with more than 20 citations per year. Since some few heavily cited items with several thousand citations a year exist—the so-called method papers and reference books—this tail of the distribution may represent a highly significant part of

Table 13.1
Source and Citation Data from SCI and SSCI

Science Citation Index

	1961	1962	1963	1964	1965	1966	1967	1968	1969	1970	1971	1972	1973	1974	1975	1976	1977
Source Journals	613	605	610	700	1146	1573	1711	1968	2180	2192	2277	2425	2364	2443	2540	2717	2655
Source Items*	113	124	129	152	236	274	304	309	341	362	364	378	407	401	419	451	495
Refs/Cites*	1370	1486	1558	1790	2925	3074	3387	3699	3850	4108	4380	4459	5017	5232	5536	6177	7398
Items Cited*	890	895	970	1092	1617	1821	1994	2139	2262	2340	2450	2597	2730	2818	3006	3246	3776
Authors Cited*	258	267	281	324	439	474	510	547	601	620	646	688	711	730	772	813	908
Cites/Item Cited	1.52	1.63	1.58	1.60	1.65	1.65	1.66	1.70	1.67	1.73	1.76	1.76	1.81	1.83	1.81	1.87	1.92
Cites/author Cited	5.23	5.67	5.44	5.38	6.07	6.36	6.51	6.52	6.28	6.52	6.57	6.65	6.95	7.05	7.05	7.68	8.01
Items/Author Cited	3.44	3.36	3.44	3.36	3.68	3.85	3.92	3.91	3.76	3.77	3.79	3.78	3.84	3.85	3.90	4.00	4.17
Refs/Source Item	12.1	12.0	12.1	11.8	12.4	11.2	11.1	12.0	11.3	11.4	12.0	12.3	12.3	13.0	13.2	13.7	14.9

*Thousands

Data from prefaces of SCI and SSCI

Social Sciences Citation Index

	1970	1971	1972	1973	1974	1975	1976
Source Journals	1000	1030	970	1052	1278	1232	1517
Source Items*	73	80	73	70	83	98	127
Refs/Cites*	618	644	604	633	872	1025	1372
Items Cited*	423	436	400	415	576	686	925
Authors Cited*	166	169	158	165	230	253	336
Cites/Item Cited	1.28	1.33	1.36	1.36	1.36	1.33	1.33
Cites/Author Cited	3.27	3.42	3.39	3.41	3.40	3.68	3.68
Items/Author Cited	2.55	2.57	2.53	2.51	2.50	2.71	2.77
Refs/Source Item	8.69	8.06	8.25	9.06	10.50	10.44	10.81

the citation behavior. Cumulative advantage theory accounts very well for the observed distribution. A fundamental parameter is the number of citations per cited paper which varies slowly, as does the number of references per source item, with the logarithm of the available archive.[6] There are now about 1.92 citations per cited item, and this is increasing linearly at 0.026 per year (see fig. 13.4). The corresponding figure for the *SSCI* is 1.33 citations per item cited, but as yet any secular increase appears to be masked by the settling down of the source selection which is still in its first few years.

As a result of this mutiplicity of citation, the 14.9S references from the S source items fall upon $14.9S/1.92 = 7.76S$ cited items. For the *SSCI* the corresponding figure is $10.8S/1.33 = 8.12S$ cited items. It should be remembered that although all source items are from journals, the cited items include also a significant proportion of books, monographs, etc. Even so, the cited items

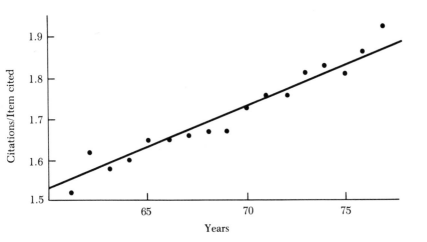

Figure 13.4.

could be only a minority of the archive available for citation, since at a growth rate of 7 percent the archive must be ca. 14S, and for the empirical growth rate of 4.14 percent for source items the archive would be 24S. Even at random, the probability of an archival item being cited at all should be in the range 0.33 to 0.57, and with a Poisson distribution the citation hits per item cited would be in the range 1.18–1.31. The significantly higher empirical figures show that cumulative advantage works very forcibly to increase the number of highly cited items beyond those that would occur with random events.

Since the cited items are sorted alphabetically by author, it is easy to make a distribution of the number of citations per cited author, or better still, the average number of cited items per cited author. At present this parameter has a value of about 4.2 for the *SCI* and 2.8 for the *SSCI*. In the former case we have enough years of data to establish a trend (see fig. 13.5); there seems to have been considerable perturbation of the parameter during the 1964–68 reorganization, but since 1969 the parameter has been

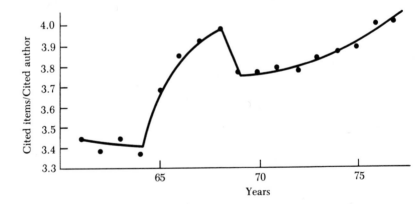

Figure 13.5.

increasing about 1 percent a year probably due more to the secular increase in collaborativeness rather than to any real increase in productivity of paper producing. Since we have 7.76 cited items in the SCI, there will be $7,76S/4.2 = 1.85S$ cited authors, and for the $SSCI$ there will be $8.12S/2.8 = 2.90S$ cited authors.

At this point in the tour of the citation cycle we may complete a loop by examining the relationship between the cited authors and the source authors. A collating of source and cited indexes shows that for both the SCI and the $SSCI$ only about half of the source authors in any year are also cited. This doubtless corresponds to the fact that about half of the year's source authors are collaborating graduate students and junior faculty without a backlog of papers of which they are the first author available for citation. The $0.55S$ first authors in the sources are therefore to be compared with the $1.85S$ that are cited in the SCI and the 2.90 cited in the $SSCI$. It follows that those active in the year are 30 percent of the SCI stock and 19 percent of the $SSCI$ stock.

Another, more accurate way of looking at the relationship is to note that we know from an independent investigation of a small slice of the SCI for a long period that only some of the collaborative authors are newcomers.[7] In fact, of the S source authors, 70 percent are continuants who publish for an extended period, and 30 percent are newcomers. Further, for the continuants we know that in any year they have a probability of 0.7 of making a publication. It follows that the S source authors imply the existence of the same number S continuants together with 0.3S newcomers. The S continuants may now be compared with the cited authors, and we derive immediately that for the SCI some $0.85/1.85$ (46 percent) and for the $SSCI$ some $1.90/2.90$ (66 percent) of the cited authors must have become discontinued by the current date. Many of the authors who once published, particularly those who published only transiently, are no longer cited; only a few are retired or deceased. It is worth noting as an overall figure that the number of cited authors in the SCI is just under a million, and in the $SSCI$ about 112,000.

Having made one circuit of the citation cycle by the comparison of source and cited authors, we may make another from the comparison of source and cited items. They have already been compared above through the medium of considering the available accrued corpus. We now look at structural relationships of the network of references to citations which, as has long been evident, knit the new layer of source papers to a small selection of highly active papers in the accrued corpus.[8] Items that are cited only once in the index are, so to speak, only tacked on to the source item that cites them, and they cannot relate two source papers or be related to any other cited paper except through this. Multiple-cited papers are comparatively rare, constituting about 27.2 percent of those in an annual index. Since we have 7.76S cited items in the SCI there must be 2.11S multiple-cited items which are connected to the S source items by about 7.63 links of reference/citation: there are therefore 7.63 links per source item and 7.63/2.11 = 3.6 links per multiple-cited item. Going to the next higher level of papers cited three or more times it turns out that the number of such papers is approximately equal to S, and the number of links at this level will be about 5.5 for each source *or* multiple-cited paper. For the $SSCI$ there is less referencing, a small corpus, and hence a lower level of multiple citation. For those papers cited twice or more we have about 1.3S which are connected to the S source papers by 4.2S links of reference/citation. These parameters enable us to establish the way in which the corpus of papers is knitted together by its links into a structure of source papers overlaying a similarly structured corpus of source papers.

A first visualization of the implied structure may be had by cutting out the very highly referencing bibliography-like sources and also the very highly cited method-like cited papers as well as those which are singly cited and cannot therefore contribute more than a tacking-on process. In this case to a first crude approximation we may suppose there to be roughly equal numbers of source and multiple-cited papers connected by about four links to and from each, respectively. We can visualize the source papers as lying on the intersections of a rectangular grid on a thin sheet which overlaps a similar grid of cited papers on a thick sheet representing several years of accretion of former sources (see fig. 13.6). Each point on the thin sheet is directly linked to the neighboring four on a complementary place of the thick sheet and vice versa. In this convention we may now see that the bibliography and method papers may be reinserted as extensive areas, rather than points, that each blanket a whole region in the other sheet (see fig. 13.7). Clearly the general form of this picture can

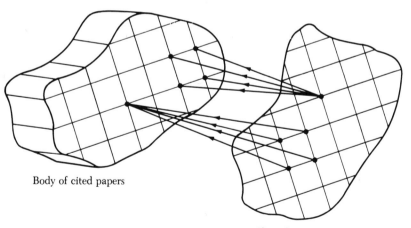

Body of cited papers

Skin of source papers

Figure 13.6.

But: —

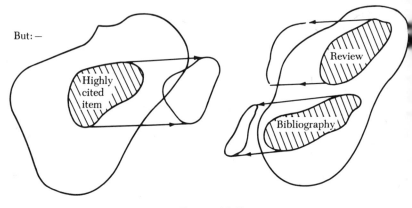

Figure 13.7.

be extended to include the moderately referencing and cited papers, too, and we may make the depiction dynamic by supposing the thick corpus of cited papers to be formed from an onion-like accretion of annual shells growing out from a nucleus laid down in the distant past.

As a next stage in this visualization we note that if there were exactly four links per item the pattern of linkage might be represented by making each intersection of a square lattice represent an item and the four lines running to it as the links. If each of the alternating source and cited items (denoted as S and C in fig. 13.8) had *exactly* four links the result would be a perfect lattice. If four is only a statistical mean, the corresponding lattice with various numbers of links would look rather like a very torn and deformable fishing net (see fig. 13.9), and if this is not envisaged in a three-dimensional analogue, the result must look rather like

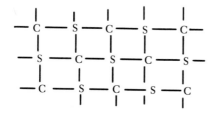

Figure 13.8.

the structure that is built into the network linkage of the corpus of science.

This property of the corpus now makes it possible to model the relational structure of what has been called "subject space."[9] It is this space that is approximately mapped by the Griffith and Small technique of cocitation analysis or that of Kessler in his bibliographic linkage which corresponds to coreferencing structure.[10] What is implied is that we have built into the citation cycle not only the quantitative modeling but also a structural scheme. In a strong sense this structure provides a natural and automatic "indexing" of the entire corpus of scientific literature, and it seems evident that many of the recall/relevance trade-off problems of actual indexing arise from a conflict between this built-in structure and that imposed by the arbitrary structure of the classifier. Not the least of the problems must be that an essentially two-dimensional skin of source papers or a three-dimensional corpus of cited papers (with time as the extra dimension) must be traversed by a classification scheme which, like the telephone book or the Dewey decimal system, is essentially a one-dimensional traversing of the map.

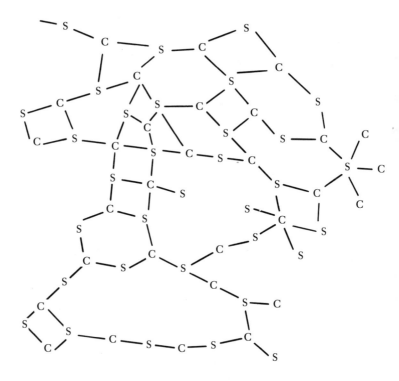

Figure 13.9.

Price's Citation Cycle
by Eugene Garfield

On a few occasions in the past, I have reprinted in *Current Contents (CC)* papers by close colleagues that seemed particularly relevant to ISI's basic mission to improve scientific communication. In the paper that follows this essay [the preceding essay of this book], my colleague and friend of twenty years, Derek Price, has written about citation analysis and the citation cycle in a particularly illuminating style.

I have always been envious of Derek's special aptitude for applying his mathematical training to scientometric problems. He has a special personality as well; he has a way of mesmerizing an audience with his Oxbridge accent, his aplomb, and his flair for the dramatic metaphor. Some scientists are irritated by his "outrageous" assertions that often come without warning, but though he sometimes seems intolerant of others' ideas, he has never been unwilling to admit that he was wrong. He is a formidable, yet affectionate, adversary.

I have often had to chastise Derek about his name and his manner of complaining about our treatment of it in the *Science Citation Index (SCI)*. As many readers probably know, his full name is Derek John de Solla Price. De Solla is a middle name, *not* a part of Derek's last name. But unless he publishes his works as Derek John de S. Price, we must index his work under De Solla-Price. Price doesn't understand why we must add the hyphen. However, there is usually no way for our indexers to know (other than personal knowledge) that his last name is just Price. Of course, his work should properly be indexed under Price, as we index under any author's last name.

Price was born January 22, 1922, and named Derek John. His father, Philip Price, was a tailor, and his mother, Fanny de Solla, a singer. Both came from early nineteenth-century Jewish immigrant families, and Derek is understandably of these origins. About 1950 he adopted his mother's Sephardic last name, de Solla, as a middle name.

Born in Leyton, a northeast London suburb, Derek was educated at British state schools. In 1938 he took a position as a physics lab assistant at the South West Essex Technical College. Since that time his work has taken him many places, including one three-year stint teaching applied mathematics at Raffles College (now the University of Singapore) in Singapore. He did war research, taught college science courses, received his B.S. in physics and mathematics in 1942, and his Ph.D. in experimental physics in 1946 (both from the University of London), before concentrating on what is now his speciality—the history of science and technology. Price also has a Ph.D. in the history of science from Cambridge University, and an honorary M.S. from Yale.

Price has had more than his share of achievements. He formulated the law of exponential growth of scientific literature, which he first presented in a paper to the VIth International Congress for the History of Science at Amsterdam in 1950.[1] This law has been a consistent basis for much of his subsequent work. He formulated the law while reading consecutive issues of the *Philosophical Transactions of the Royal Society* from 1665 to 1850. He was storing this complete set in his home for Raffles College while its library was being built, and took the opportunity to read it.

While working on a thesis for his second doctorate, Price accidentally discovered a Middle English manuscript which described the construction of a planetary calculating instrument. He identified the piece as a companion to Chaucer's 1391 *Treatise on the Astrolabe*. He then proved it to be an author's draft holograph, the only lengthy piece of Chaucer's writing known.[2]

In 1961, Price published the first of his well-known books on the history of science, *Science Since Babylon*. In it, Price began developing his idea of the exponential growth of science. Said one reviewer: "All who are interested in general history, as well as the history of science, should read and ponder upon this essay."[3] The same theme was expanded in his later book, *Little Science, Big Science*, published in 1963. Writing in the *New York Times Book Review*, John Pfeiffer, author of *The Changing Universe*, commented that Price had succeeded in this book "in bringing together considerable information on one of the outstanding phenomena of the times, the rise of science to a point where it is attracting a larger and larger proportion of our most gifted and imaginative students.[4] Both books have been highly cited— enough to classify them as classics.

Price has been extremely influential in the field of history of science at Yale University. He has been at Yale since 1959, when he was appointed Avalon professor of history of science. In addition to his duties there, he has been involved in numerous other scientific activities. He has worked with the United Nations Educational, Scientific, and Cultural Organization (UNESCO) and other agencies as a science policy advisor, and has been a consultant to many nations, among them Denmark, Germany, Switzerland, Brazil, Argentina, Israel, Egypt, Australia, India, Pakistan, the East European nations, and the Soviet Union. He helped found and served as the first president of the International Council of Scientific Unions, as international organization which promotes scientific research.

Derek has continued with his own research throughout, and has published extensively during his career—so far, more than 200 scientific papers and six books, and more are on the way. His works have covered various subjects, many of them dealing with ancient scientific instruments. One of his proudest achievements was the solution to the problem of the Greek Antikythera mechanism. This geared mechanism was discovered in 1900, and though it was of great interest to scholars, its function was unknown. Using gamma-radiology photographs of the inside of the object, Price showed it to be a sophisticated mechanical calendar previously not thought technically possible in the first century B.C.[5]

He has held more honorary posts and fellowships than I could ever hope to list here, and he has won many awards. One of the most recent was the Society for the History of Technology's Leonardo da Vinci medal. This medal is awarded annually to an individual "who has made an outstanding contribution to the history of technology by research, teaching, publication, or otherwise."[6] When the award was presented, the society had this to say about Derek: "Derek de Solla Price: You have left the mark of your researches throughout the broad field of the history of technology from the core to the periphery. At all points you have shown the skill of a virtuoso, and the depth of perception which marks the truly gifted researcher combined with the foresight which has opened new areas for others to follow." There is little that I can add.

Derek has long been associated with ISI. He has served on the editorial advisory board of the *Science Citation Index* since 1964. He also serves on the boards of the *Arts and Humanities Citation Index*, the *Social Sciences Citation Index*, *CC/Arts and Humanities*, and *CC/Social and Behavioral Sciences*. He chides us constantly for neglecting to exploit more adequately the statistical data which we generate each year in the creation of our indexes and for neglecting the more sophisticated statistics we could generate. I understand his frustration, but such desires have usually had to take second place to our production needs. He compensates partially for our failing, however, by putting our data to good use in the following paper, as he has done on numerous occasions, such as his 1965 paper "Networks of Scientific Papers."[7]

In recent years, Derek has been preoccupied with the process and theory of cumulative advantage.[8] It is cited and used in the paper which follows, and was discussed in my recent essay on Bradford's law.[9] The concept of cumulative advantage was introduced by Robert K. Merton in 1942 and later developed in his analysis of the "Matthew Effect."[10] The paper reprinted here tackles a new aspect of citation analysis, however. In this article, Derek tours the citation cycle, and discovers a built-in structure to citation relationships. He provides a model for this structure that can help the student of citation analysis visualize the various dimensions and interworkings of the citation cycle. I am pleased to be able to reprint this important contribution to the study of science, and pleased that it gave me a chance to offer this small tribute to my great friend Derek Price.

Notes

1. Prologue to a Science of Science

1. Gerald Holton, "On the recent past of physics," *American Journal of Physics* (December 1961), 29:805. I should like to draw attention to the fine study published while this work was in progress: Gerald Holton, "Models for Understanding the Growth and Excellence of Scientific Research," in S. R. Graubard and G. Holton, eds., *Excellence and Leadership in a Democracy*, pp. 94–131 (New York: Columbia University Press, 1962), first published as "Scientific research and scholarship: Notes towards the design of proper scales," in *Proceedings of the American Academy of Arts and Sciences*, 91(2):362–99 (*Daedalus*, March 1962). My work derives much from this previous publication, though its author and I do not always agree in detail in the conclusions we derive from the statistical data.

2. Alvin M. Weinberg, "Impact of large-scale science on the United States," *Science* (July 21, 1961), 134:164. I am indebted to this paper for many ideas. See also further comments by Weinberg in "The Federal Laboratories and science education," *Science* (April 6, 1962), 136:27.

3. It is easy enough to convert from one to the other by noting, as a rough approximation, that 10 doubling periods correspond to a factor of 1024, or about 3 tenfolding periods.

4. For a more detailed discussion of this see Derek J. de Solla Price, *Science Since Babylon* (New Haven: Yale University Press, 1961), chapter 5.

5. An excellent historical account of the birth of scientific journals is given by David A. Kronick, *A History of Scientific and Technical Periodicals* (New York: Scarecrow Press, 1962).

6. S. G. Lasky, "Mineral industry futures can be predicted," *Engineering and Mining Journal* (August 1951), 152:60; (September 1955), 156:94.

2. Galton Revisited

1. Karl Pearson, *The Life, Letters, and Labours of Francis Galton* (New York: Cambridge University Press, 1914–1930), see especially vol. 3a, p. 125.

2. Francis Galton, *Hereditary Genius* (London, 1869; reprinted by Meridian Books, 1962).

3. Here and later I have made considerable use of the extensive analysis in S. S. Visher, *Scientists Starred 1903–1943* in American Men of Science, (Baltimore: Johns Hopkins University Press, 1947).

4. The exception being Riemann, who published only 19 papers but died at the age of forty.

5. Productivity is therefore one of many factors.

6. Alfred J. Lotka, "The frequency distribution of scientific productivity," *Journal of the Washington Academy of Sciences*, (1926) 16:317. For a fuller analysis and justification see Herbert A. Simon, *Models of Man, Social and National* (New York: Wiley, 1957), p. 160.

7. Wayne Dennis, "Bibliographies of eminent scientists," *Scientific Monthly* (September 1954), 79:180–83.

8. For the graphical presentations given here we have used a distribution law of the form

$$N = k\left(\frac{1}{p} - \frac{1}{a + p}\right) = \frac{ak}{p(a + p)},$$

where N is the cumulative number of men who publish at least p scientific papers within a given interval of time (here taken as a lifetime). For those of low productivity $1 < p < a$, and the law tends to the inverse first-power from $N = k/p$, while for the high-productivity authors we have $a < p < \infty$, and the law approximates the inverse-square form $N = ak/p^2$. We find that the available data may be fitted by taking the boundary between high and low productivities at $a = 15$ papers per lifetime.

From the given distribution law one is able to compute in sequence the number of people publishing exactly p papers, the number of papers published in all by such people and, finally, the cumulative number of papers published by the cumulative number of authors. This enables one to calculate all the properties of such a distribution in terms of the parameter, a, and the arbitrary constant of proportionality, k. It happens, for example, that the average number of papers per author is given by $1 + (1 + 1/a) \log (1 + a)$ which is very insensitive to the magnitude of a, assuming a value of 3 for $a = 7$, and a value of 4 for $a = 22$.

9. The law was proposed in detail in V. Pareto, *Cours d'économie politique* (1897), 2: 299–345.

10. Cf. Galton's citation of the marks gained by the wranglers in the Cambridge Mathematical Tripos. Their scores, as nearly as possible on an objective open scale of merit, were such that the top candidate in each year got almost twice the marks of the second, and thirty times that of the 100th candidate.

The log-normal character of scientific productivity distributions has previously been suggested by William Shockley, "On the statistics of individual variations of productivity in research laboratories," *Proceedings of the Institute of Radio Engineers* (1957), 45:279,1409.

11. Lindsey R. Harmon, "The high school backgrounds of science doctorates," *Science* (March 10, 1961), 133:679, also published at length in *Scientific Manpower 1960* (NSF 61-34, May 1961), pp. 14–28.

12. This is more than twice the present *world* population of scientists.

13. This, then, provides a measure that is linear, not exponential. It is the sort of index which might correspond with Nobel Prizes (which come linearly with time because that is how they are organized) and possibly also with unexpected, crucial advances.

14. Figure 2.4 and the following data are from George K. Zipf, *Human Behavior and the Principle of Least Effort* (Cambridge, Mass.: Addison-Wesley, 1949), p.420, figure 10-2.

15. Data from a preliminary survey of scientific periodicals by the Library of Congress.

3. Invisible Colleges and the Affluent Scientific Commuter

1. However, publications of the learned academies as corporate bodies engaged in the experiments and trials for which they had been constituted had appeared before. The *Saggi* of the Accademia del Cimento, which preceded the societies of London and Paris, is a volume of fine research papers published as a complete and final single book, not as a serial.

2. Bernard Barber, "Resistance by scientists to scientific discovery," *Scientific Manpower 1960* (NSF 61-34, May 1961), pp. 36–47.

3. Robert K. Merton, "Priorities in scientific discovery: A chapter in the sociology of science," *American Sociological Review*, (1957), 22:635; "Singletons and multiples in scientific discovery: A chapter in the sociology of Science," *Proceedings of the American Philosophical Society* (October 1961), 105:470.

4. Thomas S. Kuhn, "Historical structure of scientific discovery,"*Science* (June 1, 1962), 136:760.

5. Cited in Merton, *"Singletons and Multiples,"* p.483.

6. F. Reif, "The competitive world of the pure scientist," *Science* (December 15, 1961),134:1957–62.

7. A beginning for such analysis has been made by Karl W. Deutsch, "Scientific and Humanistic Knowledge in the Growth of Civilization," in *Science and the Creative Spirit* (Toronto: University of Toronto Press, 1958), pp. 3–51.

8. One may, however, *scan* a group of this size or even larger—using, say, one of the abstrasts journals in order to find the small group.

9. One might well look into the motivation of such founders. Compare the story in which two little girls seize control of their fourth-grade discussion club by a method that they described as "the fair and square way by which any group takes over any club—capture of the mimeograph machine."

10. D. J. Urquhart, "Use of Scientific Periodicals," International Conference on Scientific Information, National Academy of Sciences–National Research Council, Washington, D.C., 1958, pp. 277–90, tables 2, 7.

11. At this point most scientists will express disappointment. I suspect they have a secret hope that some standard will be found for the objective judgment of their own caliber and reputation. This craving for a recognition unsullied by human subjectivity is in itself an interesting psychological phenomenon.

12. Papers behave rather like a human population, except that it seems to take

a quorum of about 10 papers to produce a new one, rather than a pair of male and female. We have now shown that childbearing proceeds at constant rate.

13. See, for example, J. H. Westbrook, "Identifying significant research," *Science* (October 28, 1960) 132:1229–34. Also Paul Weiss, "Knowledge, a growth process," *Science* (June 3, 1960), 131:1716, and the clarifying subsequent discussion by S. J. Goffard and C. D. Windle, *Science* (September 2, 1960), 132:625.

14. Herman H. Fussler, "Characteristics of the research literature used by chemists and physicists in the United States," *Library Quarterly* (1949), 19:19–35; (1950), 20:119–143.

15. P. L. K. Gross and E. M. Gross, "College libraries and chemical education," *Science* (October 28, 1927), 66:385–89.

16. In fact a constant rate of citation will ensure that the field increases with compound interest so that the growth is exponential.

17. I therefore arrive at the conclusion that a scientific race to get there first is tremendously wasteful, and that anything that lessens the reward for such achievement is good. Thus it is perhaps a good thing to deprive the authors of their chance to get their names on the paper. It might be made sufficient honor and reward that they are allowed to play with the team.

18. Like government contract research reports, these represent an obnoxious (though historically interesting) backdoor means of getting publication for a mass of writing that might be better lost.

19. S. A. Goudsmit, *Physical Review Letters* (March 15, 1962), 8:229. Another good example of a quite different sort of collaboration is the appearance of the world's greatest pseudonymous mathematician, Nicolas Bourbaki. This Frenchman with a Greek name, author of an internationally famous collection of treatises on modern higher mathematics, is actually a group of ten to twenty mathematicians, most of them French, all of them highly eminent in their fields, none of them identified by name as part of the polycephalic Bourbaki. See Paul R. Halmos, "Nicolas Bourbaki," *Scientific American* (May 1957), 196:88–99.

20. Results of an unpublished investigation by L. Badash, Yale University.

21. L. Kowarski, "Team work and individual work in research," *CERN Courier* (May 1962), 2:4–7.

22. Cf. data for *Mathematical Reviews* and three United States mathematics journals (percent papers having joint authors):

	Math. Revs.	Three U.S. journals
1920		2.2
1930		4.1
1940	5.8	18.2
1950	6.5	18.2
1960	10.8	12.7

From a letter by W. R. Utz, *American Mathematical Society Notices*, (1962), 9:196–97.

23. Letter in *Physics Today* (June 1962), 15:79–80.

4. Political Strategy for Big Scientists

1. Dale R. Lindsay and Ernest M. Allen, "Medical research: Past support, future directions," *Science* (December 22, 1961), 134:2017–24.

2. Reported editorially in *Science* (August 26, 1960), 132:517.

3. I am grateful to Eri Yagi Shizume, Yale University, for allowing me to make use of her data, to be published in *Proceedings XII International Congress for the History of Science*, Ithaca and Philadelphia, August 1962.

4. USSR   1950   approx.   500 journals
           1960   1500

   China   1949   0
           1959   400

5. For a masterly and heartfelt analysis of this problem see Stevan Dedijer, "Why did Daedalus leave?" *Science* (June 30, 1961), 133:2047–52.

6. The increase of status is analyzed in Bentley Glass, "The academic scientist, 1940–1960," *Science* (September 2, 1960), 132:598–603.

7. The term is used by anthropologists to describe the reactions of primitive peoples to boatloads of civilization. In the Pacific Isles, in the last war, when the Navy arrived the native huts were decked with bamboo facsimiles of radar antennae, put there so that the new gods would smile on them and bring riches. (Story told by A. Hunter Dupree, "Public education for science and technology," *Science* [September 15, 1961], 134:717.)

8. Bernard Barber, "Resistance by scientists to scientific discovery," *Science* (September 1, 1961), 134:596–602.

9. Donald W. MacKinnon, "What makes a person creative?" *Saturday Review* (February 10, 1962), 45:15–17, 69; "The Nature and Nurture of Creative Talent," Bingham Lecture, Yale University, April 11, 1962.

10. Newton, Kelvin, Lavoisier, Boyle, Huygens, Count Rumford, Mme. Curie, and Maxwell are examples.

11. Very little reliable work has been published on the psychology of scientists. The only books known to me are Anne Roe, *The Making of a Scientist* (New York: Apollo Editions, reprint A-23) and Bernice T. Eiduson, *Scientists: Their Psychological World* (New York: Basic Books, 1962).

12. According to a survey by Science Service reported editorially in *Science* (September 30, 1960), 132:885.

13. Robert Gilpin, *American Scientists and Nuclear Weapons Policy* (Princeton: Princeton University Press, 1962).

5. Networks of Scientific Papers

1. Eugene Garfield and I. H. Sher, "New factors in the evaluation of scientific literature through citation indexing," *American Documentation* (1963), 14:191; Garfield and Sher, *Genetics Citation Index* (Philadelphia: Institute for Scientific Information, 1963). For many of the results discussed in this essay I have used

statistical information drawn from Garfield and Sher, *Science Citation Index* (Philadelphia: Institute for Scientific Information, 1963), pp. ix, xvii–xviii.

I wish to thank Dr. Eugene Garfield for making available to me several machine printouts of original data used in the preparation of the 1961 *Index* but not published in their entirety in the preamble to the index.

I am grateful to Dr. M. M. Kessler, Massachusetts Institute of Technology, for data for seven research reports of the following titles and dates: "An Experimental Study of Bibliographic Coupling Between Technical Papers" (November 1961); "Bibliographic Coupling Between Scientific Papers" (July 1962); "Analysis of Bibliographic Sources in the *Physical Review* (vol. 77, 1950, to vol. 112, 1958)" (July 1962); "Analysis of Bibliographic Sources in a Group of Physics-related Journals" (August 1962); "Bibliographic Coupling Extended in Time: Ten Case Histories" (August 1962); "Concerning the Probability that a Given Paper will be Cited" (November 1962); "Comparison of the Results of Bibliographic Coupling and Analytic Subject Indexing" (January 1963).

J. W. Tukey, "Keeping research in contact with the literature: Citation indices and beyond," *Journal of Chemical Documentation*, (1962), 2:34.

C. E. Osgood and L. V. Xbignesse, *Characateristics of Bibliographical Coverage in Psychological Journals Published in 1950 and 1960* (Urbana: Institute of Communications Research, University of Illinois, 1963).

R. E. Burton and R. W. Kebler, "The 'half-life' of some scientific and technical literatures, *American Documentation* (1960), 11:18.

6. Collaboration in an Invisible College

1. We wish to thank the members of this group, particularly its chairman, David E. Green, Institute for Enzyme Research, University of Wisconsin, and Errett C. Albritton, Director, IEG Program, National Institutes of Health, for permitting us this research access and providing a file of research memos. This research has been supported by Grant GN–299 (continued as GN–527) from the National Science Foundation.

2. For a history of IEG see David E. Green, "An experiment in communications: The information exchange group," *Science* (1964), 153:308–9, and "Information exchange group number one," *Science* (1965), 148:1543; see also two unpublished accounts by Errett C. Albritton: "The Information Exchange Group—an Experiment in Communication" (July 1965), and "The Information Exchange Group—An Experimental Program" (February 1966). Another account with bibliography is given by C. A. Moore, "An old information device with new outlooks," *Journal of Chemical Documentation* (1965), 5(3):126–28.

3. See, for example, M. J. Moravcsik, "On improving communication," *Bulletin of the Atomic Scientists* (1966), 22(5):31, and S. Pasternack, "Is journal publication obsolescent?" *Physics Today* (1966), 19(5):38–43.

4. It consists of memos numbered 1 through 535 with the exception of the following 11 [sic] which were not circulated or were otherwise unavailable to us: 36, 77, 257, 263, 308, 321, 400, 412, 491, and with the addition of 9 papers numbered irregularly as X34, X35, X36, Y36, Sp. 1, Sp. 1a, 363A, X34, 273–II.

5. Beverly L. Clarke, "Multiple authorship trends in scientific papers," *Science* (1964), 143:822–24.

6. Stella Keenan and Pauline Atherton, *The Journal Literature of Physics* (New York: American Institute of Physics, AIP/DRP PA1, 1964), p. 5.

7. Walter Hirsch and James F. Singleton, "Research support, multiple authorship, and publication in sociological journals, 1936–1964," preprint, June 1965.

8. To avoid personal issues we identify each author not by name but by a two-letter code acronym.

7. Measuring the Size of Science

1. Derek J. de Solla Price "The Science of Science," in M. Goldsmith and A. Mackay, eds., *The Science of Science*, (New York: Souvenir Press, 1964).

2. Robert K. Merton, "Priorities in scientific discovery—a chapter in the sociology of science," *American Sociological Review* (1957), 22:635.

3. J. D. Reuss, *Repertorium Commentationum a Societatibus Litterariis Editarum, vols. 1–16, (Göttingen, 1805).

4. Stella Keenan and Pauline Atherton, *The Journal Literature of Physics* (New York: American Institute of Physics, AIP/DRP PA1, 1964).

5. Price, "Nations can publish or perish," *International Science and Technology* (1967), 70:84–90.

6. Note that this figure of 1 percent is not constant. It must double every 10 years or thereabouts, so that it will become 2 percent by the mid–1970s. Scientific research grows in manpower, and it becomes more expensive per man, and both of these rates of growth are at least as rapid as that of the GNP.

8. Citation Measures of Hard Science, Soft Science, Technology, and Nonscience

This work was supported by National Science Foundation grants GN–299 and GN–527. I am grateful to the Guggenheim Foundation for providing me the sabbatical leisure to effect some synthesis of results previously obtained. I would also like at this time to acknowledge the research collaboration for several years of Diana Crane, and a special debt to the pioneer labors of Robert Merton, doyen of the sociology of science. It is a pleasure, too, to be able to give special thanks to Eugene Garfield and his colleagues at the Institute for Scientific Information, Philadelphia, for without their free provision of data from the *Science Citation Index* and other publications and services this work would have been impossible.

1. Edwin B. Parker, William J. Paisley, Roger Garrett, *Bibliographic Citations As Unobtrusive Measures of Scientific Communication* (Stanford: Stanford University Institute for Communication), October 1967.

2. Karl W. Deutsch, "Scientific and Humanistic Knowledge in the Growth of Civilization," in Harcourt Brown, ed., *Science and the Creative Spirit*, pp. 3–51 (Toronto: University of Toronto Press, 1958).

3. Norman W. Storer, "The hard sciences and the soft: Some sociological observations," *Bulletin of the Medical Library Association* (1967), 55:75–84.

4. Kenneth M. Wilson, "Of Time and the Doctorate," *Southern Regional*

Education Board Research Monograph No. 9 (Atlanta, 1965). See especially table 1.2, p. 14.

5. Richard H. Bolt, Walter L. Koltrun, and Oscar H. Levine, "Doctoral feedback into higher education," *Science* (May 14, 1965), 148:918–28.

6. Warren O. Hagstron, *The Scientific Community* (New York: Basic Books, 1965).

7. It seems to me a great pity to waste a good technical term by using the words *citation* and *reference* interchangeably. I therefore propose and adopt the convention that if Paper R contains a bibliographic footnote using and describing Paper C, then R contains a reference to C, and C has a *citation* from R. The number of references a paper has is measured by the number of items in its bibliography as endnotes and footnotes, etc., while the number of citations a paper has is found by looking it up on some sort of citation index and seeing how many other papers mention it.

8. See essay 6.

9. Walter Hirsch and James F. Singleton, "Research Support, Multiple Authorship, and Publications in Sociological Journals, 1936–1964," unpublished report, Purdue University, 1964.

10. See essay 5.

11. Duncan MacRae, Jr., "Growth and decay curves in scientific citations," *American Sociological Review* (1969), 34(5):631–35.

12. *Ibid.*

13. D. R. Stoddart, "Growth and structure of geography," *Transactions and Papers of the Institute of British Geographers* (June 1967), 41:1–19.

14. R. E. Burton and R. W. Kebler, "The half-life of some scientific and technical literature," *American Documentation* (1960), 11:18–22.

15. A. J. Meadows, "The citation characteristics of astronomical research literature," *Journal on Documentation* (March 1967), 23:28–33.

16. E. R. N. Grigg, "Information science and American radiology," *Radiologic Technology* (1969), 41:19–30.

17. Parker, Paisley, and Garrett, *Bibliographic Citations* .

18. It is a pleasure to record my thanks to Mrs. Janet Williams for this efficient piece of noncomputerized counting.

9. Some Statistical Results for the Numbers of Authors in the States of the United States and the Nations of the World

We express our thanks to Professor Francis Anscombe, Department of Statistics, Yale University, for most helpful advice on statistical computing programs and methodologies. This work was performed under a National Science Foundation Research Grant #SOC 73–05428.

1. See essay 7.

10. Studies in Scientometrics, Part 1: Transience and Continuance in Scientific Authorship

1. We should like to express our deep gratitude to Dr. Eugene Garfield and the officers and staff of ISI for their very effective long-standing cooperation and technical help in this and related projects. The investigation was supported under Grant GS–39830X from the National Science Foundation.

2. A full account of these laws and reference to the relevant literature is given earlier in this book in essay 2.

13. The Citation Cycle

1. Another advantage: this essay acknowledges no support whatsoever from any agency or foundation, but then, no time wasted, either, from preparing and submitting proposals.

2. For a general survey of the bibliometrics of citation see Francis Narin, *Evaluative Bibliometrics*, Computer Horizons, Inc., Project no. 704R, March 31, 1976; Eugene Garfield, *Citation Indexing: Its Theory and Application in Science, Technology, and Humanities* (New York: Wiley, 1979); and Roland Hjerppe, *An Outline of Bibliometrics and Citation Analysis* (Stockholm: Royal Institute of Technology Library), October 1978.

3. Derek J. de Solla Price, "A general theory of bibliometric and other cumulative advantage processes," *Journal of the American Society for Information Science* (1976), 27(5/6):292–306; and "Cumulative advantage urn games explained: A reply to Kantor," *Journal of the American Society for Information Science*, (1978), 29(4):204–6. For a recent large-scale empirical test of the Bradford approximation see M. Carl Drott and Belver C. Griffith, "An empirical examination of Bradford's law and the scattering of scientific literature," *Journal of the American Society for Information Science* (September 1978), 29:238–46.

4. See essay 6.

5. For a history of scientific collaboration see Donald deB. Beaver and R. Rosen, "Studies in scientific collaboration, Part I: The professional origins of scientific co-authorship," *Scientometrics* (1978), 1:65–84; "Part II: Scientific co-authorship, research productivity, and visibility in the French scientific elite, 1799–1830," *Scientometrics* (1979), 1:133–49; and "Part III: Professionalization and the natural history of modern scientific co-authorship," *Scientometrics* (1979), 1:231–45.

6. Eugene Garfield, "Is the ratio between number of citations and publications cited a true constant?" *Current Contents* (February 9, 1976), vol. 6, editorial.

7. See essay 11.

8. See essay 5.

9. Peter P. M. Meincke and Pauline Atherton, "Knowledge space: A conceptual basis for the organization of knowledge," *Journal of the American Society for Information Science* (January-February 1976), 27:18–24, and Michael J. McGill,

286 PRICE'S CITATION CYCLE

"Knowledge and information spaces: Implications for retrieval systems," *Journal of the American Society for Information Science* (July-August 1976), 27:205–10.

10. H. G. Small and Belver C. Griffith, "The structure of scientific literature 1: Identifying and graphing specialties," *Science Studies* (1974), 4:17–40; Belver C. Griffith and H. G. Small, "The structure of scientific literature II: The macro- and micro-structure of science," *Science Studies* (1974), 4:339–65; H. G. Small, "A co-citation model of a scientific specialty: A longitudinal study of collagen research," *Social Studies of Science* (1977), 7:139–66; and H. G. Small and Edwin Greenlee, "Citation context analysis of a co-citation cluster: Recombinant-DNA," *Scientometrics* (1980), 2(4):277–301.

Price's Citation Cycle

My thanks to Susan Fell Evans for her help in the preparation of this essay.

1. Derek J. de Solla Price, "Quantitative measures of the development of science," *Archives Internationales d'Histoire des Sciences* (1951), 14:85–93; *Science Since Babylon* (New Haven: Yale University Press, 1975); *Little Science, Big Science*, 1963 edition.

2. Price, *The Equatorie of the Planetis* (Cambridge: Cambridge University Press, 1955).

3. D. Fleming, *Science Since Babylon* (book review), American Historical Review (1962), 68:170.

4. John Pfeiffer, "Problems raised to a higher level," *New York Times Book Review*, June 16, 1963, p. 6.

5. Price, "Antikythera mechanism," *Proceedings of the Fourteenth International Congress of the History of Science*, August 19–27, 1974, Tokyo and Kyoto, Japan (Tokyo: Science Council of Japan, 1975), p. 193–96.

6. "The Leonardo da Vinci medal," *Technology and Culture* (1977), 18:471–78.

7. See essay 5.

8. Price, "A general theory of bibliometric and other cumulative advantage processes," *Journal of the American Society for Information Science* (1976), 27:292–306; "Cumulative advantage urn games explained: a reply to Kantor," *Journal of the American Society for Information Science* (1978), 29(4):204–6.

9. Eugene Garfield, "Bradford's law and related statistical patterns," *Current Contents* (May 12, 1980), 19:5–12.

10. Robert K. Merton "The Normative Structure of Science," in Merton, *The Sociology of Science*, Norman W. Storer, ed. (Chicago: University of Chicago Press, 1973), p. 221–78.

Index